Communications
in Computer and Information Science

2197

Editorial Board Members

Joaquim Filipe ⓘ, *Polytechnic Institute of Setúbal, Setúbal, Portugal*
Ashish Ghosh ⓘ, *Indian Statistical Institute, Kolkata, India*
Lizhu Zhou, *Tsinghua University, Beijing, China*

Rationale

The CCIS series is devoted to the publication of proceedings of computer science conferences. Its aim is to efficiently disseminate original research results in informatics in printed and electronic form. While the focus is on publication of peer-reviewed full papers presenting mature work, inclusion of reviewed short papers reporting on work in progress is welcome, too. Besides globally relevant meetings with internationally representative program committees guaranteeing a strict peer-reviewing and paper selection process, conferences run by societies or of high regional or national relevance are also considered for publication.

Topics

The topical scope of CCIS spans the entire spectrum of informatics ranging from foundational topics in the theory of computing to information and communications science and technology and a broad variety of interdisciplinary application fields.

Information for Volume Editors and Authors

Publication in CCIS is free of charge. No royalties are paid, however, we offer registered conference participants temporary free access to the online version of the conference proceedings on SpringerLink (http://link.springer.com) by means of an http referrer from the conference website and/or a number of complimentary printed copies, as specified in the official acceptance email of the event.

CCIS proceedings can be published in time for distribution at conferences or as post-proceedings, and delivered in the form of printed books and/or electronically as USBs and/or e-content licenses for accessing proceedings at SpringerLink. Furthermore, CCIS proceedings are included in the CCIS electronic book series hosted in the SpringerLink digital library at http://link.springer.com/bookseries/7899. Conferences publishing in CCIS are allowed to use Online Conference Service (OCS) for managing the whole proceedings lifecycle (from submission and reviewing to preparing for publication) free of charge.

Publication process

The language of publication is exclusively English. Authors publishing in CCIS have to sign the Springer CCIS copyright transfer form, however, they are free to use their material published in CCIS for substantially changed, more elaborate subsequent publications elsewhere. For the preparation of the camera-ready papers/files, authors have to strictly adhere to the Springer CCIS Authors' Instructions and are strongly encouraged to use the CCIS LaTeX style files or templates.

Abstracting/Indexing

CCIS is abstracted/indexed in DBLP, Google Scholar, EI-Compendex, Mathematical Reviews, SCImago, Scopus. CCIS volumes are also submitted for the inclusion in ISI Proceedings.

How to start

To start the evaluation of your proposal for inclusion in the CCIS series, please send an e-mail to ccis@springer.com.

Ludovico Boratto · Daniele Malitesta ·
Mirko Marras · Giacomo Medda ·
Cataldo Musto · Erasmo Purificato
Editors

Advances on Graph-Based Approaches in Information Retrieval

First International Workshop, IRonGraphs 2024
Glasgow, UK, March 24, 2024
Proceedings

Editors
Ludovico Boratto ⓘ
University of Cagliari
Cagliari, Italy

Mirko Marras ⓘ
University of Cagliari
Cagliari, Italy

Cataldo Musto ⓘ
University of Bari Aldo Moro
Bari, Italy

Daniele Malitesta ⓘ
Polytechnic University of Bari
Bari, Italy

Giacomo Medda ⓘ
University of Cagliari
Cagliari, Italy

Erasmo Purificato ⓘ
Otto von Guericke University
Magdeburg, Germany

ISSN 1865-0929 ISSN 1865-0937 (electronic)
Communications in Computer and Information Science
ISBN 978-3-031-71381-1 ISBN 978-3-031-71382-8 (eBook)
https://doi.org/10.1007/978-3-031-71382-8

© The Editor(s) (if applicable) and The Author(s), under exclusive license
to Springer Nature Switzerland AG 2025

This work is subject to copyright. All rights are solely and exclusively licensed by the Publisher, whether the whole or part of the material is concerned, specifically the rights of translation, reprinting, reuse of illustrations, recitation, broadcasting, reproduction on microfilms or in any other physical way, and transmission or information storage and retrieval, electronic adaptation, computer software, or by similar or dissimilar methodology now known or hereafter developed.
The use of general descriptive names, registered names, trademarks, service marks, etc. in this publication does not imply, even in the absence of a specific statement, that such names are exempt from the relevant protective laws and regulations and therefore free for general use.
The publisher, the authors and the editors are safe to assume that the advice and information in this book are believed to be true and accurate at the date of publication. Neither the publisher nor the authors or the editors give a warranty, expressed or implied, with respect to the material contained herein or for any errors or omissions that may have been made. The publisher remains neutral with regard to jurisdictional claims in published maps and institutional affiliations.

This Springer imprint is published by the registered company Springer Nature Switzerland AG
The registered company address is: Gewerbestrasse 11, 6330 Cham, Switzerland

If disposing of this product, please recycle the paper.

Advances on Graph-Based Approaches in Information Retrieval: Preface

The First International Workshop on Graph-Based Approaches in Information Retrieval (IRonGRAPHS 2024) was held as part of the 46th European Conference on Information Retrieval (ECIR 2024) on March 24, 2024. IRonGRAPHS 2024 was held in Glasgow, Scotland, with support for remote attendance. The workshop was jointly organized by University of Cagliari (Italy), Polytechnic University of Bari (Italy), University of Bari (Italy), and Otto von Guericke University Magdeburg (Germany). This year, the workshop counted 14 submissions from different countries. All submissions were double-blind peer-reviewed by at least three internal Program Committee members, ensuring that only high-quality work was then included in the final workshop program. The pool of reviewers integrated both new and accomplished researchers in the field from industry and academia. The final program included 6 full papers.

The workshop day included an interesting paper session, an engaging spotlight session, and a final discussion to highlight open issues and research challenges, and summarize the workshop's outcomes. The paper session showcased diverse novel contributions, with presentations on knowledge-aware graph-based recommender systems using user-based semantic features filtering, source-target node distance impacts on adversarial attacks in social network recommendations, the effectiveness of graph contrastive learning for mathematical information retrieval, innovative methods for identifying user shopping needs through voice product questions, the application of graph-based retrieval-augmented generation in soccer data analysis, and enhanced semantic understanding with graph-based information retrieval. The spotlight session featured rapid presentations on emerging research areas, including legal case retrieval with text-attributed graphs, explainable keyword search over knowledge graphs, session-based recommendation algorithms based on graph neural networks, query expansion for event-specific document ranking, hypergraphs with attention on reviews for explainable recommendations, and recall failures in ad hoc search using graphs. The day concluded with a final discussion and closing remarks.

In addition to the paper presentations, the program also included two keynote talks. As for the first one, Francesco Fabbri from Spotify (Spain) emphasized the importance of integrating consumption signals and content-based representations through graph-based learning methods, demonstrating how these methods combined with large language models can enhance personalization at scale for various downstream tasks in the digital audio content domain. In the other talk, Ruihong Qiu from the University of Queensland (Australia) highlighted how graph learning captures structural characteristics, converting user history and legal cases into graphs to integrate insights into user behavior and legal context, demonstrating the promise of leveraging graph data for valuable information retrieval. These examples showed the potential of graph learning techniques.

The first edition of the workshop was a success, with a consistent level of engagement throughout. IRonGRAPHS 2024 strengthened the community working on graph-based approaches in information retrieval, representing a key event where ideas and solutions for current challenges are discussed. This success motivates us to organize the second edition of the workshop next year. The organizers would like to thank the authors and reviewers for shaping an interesting program, and the attendees for their enthusiastic participation.

July 2024

Ludovico Boratto
Daniele Malitesta
Mirko Marras
Giacomo Medda
Cataldo Musto
Erasmo Purificato

Organization

Program Committee Chairs

Ludovico Boratto	University of Cagliari, Italy
Daniele Malitesta	Polytechnic University of Bari, Italy
Mirko Marras	University of Cagliari, Italy
Giacomo Medda	University of Cagliari, Italy
Cataldo Musto	University of Bari, Italy
Erasmo Purificato	Otto von Guericke University Magdeburg, Germany

Program Committee

Abiola Akinnubi	University of Arkansas at Little Rock, USA
Manuel Dileo	University of Milan, Italy
Vittoria Cozza	ENEA, Italy
Tendai Mukande	Dublin City University, Ireland
Alessandro De Bellis	Polytechnic University of Bari, Italy
Ladislav Peska	Charles University, Czechia
Emanuel Lacic	Infobip, Croatia
Julia Neidhardt	TU Wien, Austria
Maurizio Ferrari Dacrema	Polytechnic University of Milan, Italy
Salvatore Bufi	Polytechnic University of Bari, Italy
Alireza Javadian Sabet	University of Pittsburgh, USA
Alejandro Bellogin	Universidad Autónoma de Madrid, Spain
Claudio Pomo	Polytechnic University of Bari, Italy
Antonela Tommasel	ISISTAN & CONICET-UNICEN, Argentina
Eva Zangerle	University of Innsbruck, Austria
Claudio Di Sipio	University of L'Aquila, Italy
Alberto Carlo Maria Mancino	Polytechnic University of Bari, Italy
Dominik Kowald	Know-Center, Austria
Rishi Bhatia	Walmart, USA
Jeongwhan Choi	Yonsei University, South Korea

Danilo Dessì	GESIS Leibniz In. for Social Sciences, Germany
Monika Shrivastav	Walmart, USA
Mukul Singh	Microsoft, USA
Edoardo D'Amico	University College Dublin, Ireland
Giacomo Balloccu	University of Cagliari, Italy

Contents

Soccer-GraphRAG: Applications of GraphRAG in Soccer 1
Zahra Sepasdar, Sushant Gautam, Cise Midoglu, Michael A. Riegler, and Pål Halvorsen

Enhanced Semantic Understanding with Graph-Based Information Retrieval 11
Giovanni M. De Filippis, Antonio M. Rinaldi, Cristiano Russo, and Cristian Tommasino

Identifying Shopping Intent in Product QA for Proactive Recommendations 25
Besnik Fetahu, Nachshon Cohen, Elad Haramaty, Liane Lewin-Eytan, Oleg Rokhlenko, and Shervin Malmasi

KGUF: Simple Knowledge-Aware Graph-Based Recommender with User-Based Semantic Features Filtering 41
Salvatore Bufi, Alberto Carlo Maria Mancino, Antonio Ferrara, Daniele Malitesta, Tommaso Di Noia, and Eugenio Di Sciascio

The Effectiveness of Graph Contrastive Learning on Mathematical Information Retrieval 60
Pei-Syuan Wang and Hung-Hsuan Chen

The Impact of Source-Target Node Distance on Vicious Adversarial Attacks in Social Network Recommendation Systems 73
Federico Albanese, Giovanni Trappolini, Lorenzo Scarlino, and Fabrizio Silvestri

Author Index 89

Soccer-GraphRAG: Applications of GraphRAG in Soccer

Zahra Sepasdar[1,2](✉)[iD], Sushant Gautam[1,3][iD], Cise Midoglu[1][iD], Michael A. Riegler[1,3][iD], and Pål Halvorsen[1,2,3][iD]

[1] SimulaMet, Oslo, Norway
zasep3162@oslomet.no
[2] Forzasys, Oslo, Norway
[3] OsloMet, Oslo, Norway

Abstract. In the realm of soccer analytics, the need for efficient and accurate information retrieval is crucial. In this paper, we introduce SoccerGraphRAG, a framework designed to facilitate the retrieval of soccer-related information through natural language queries. This system leverages knowledge graphs, created from the recently released SoccerNetEchoes dataset which includes transcriptions of soccer game audio commentaries. Soccer-GraphRAG aims to streamline the retrieval, access, and analysis of soccer data, providing insights with precision and contextual relevance. This framework is ideally suited for analyzing player performance, as well as for engaging in question answering (Q&A) and summarizing tasks.

Keywords: Association Football · Knowledge Graphs · GraphRAG · Automatic Speech Recognition (ASR) · Large Language Models (LLM)

1 Introduction

In recent years, Large Language Models (LLMs) have made significant strides in understanding and generating human-like text [1,14]. Despite these advances, LLMs face critical limitations that hinder their reliability and applicability in practical scenarios. One major issue is the phenomenon known as 'hallucinations', where models generate plausible but factually incorrect or nonsensical information [1,3,4]. Additionally, LLMs are constrained by the static nature of their training data, limiting their ability to incorporate the most recent information or answer queries outside their training corpus [3]. To address these limitations, the Retrieval Augmented Generation (RAG) approach offers a compelling solution. By integrating external data dynamically at query time, RAG enhances the ability of LLMs to produce more accurate and contextually relevant responses [3,9]. However, traditional RAG methods relying on vector search cannot capture the full complexity of information [7]. Incorporating a dynamic knowledge base with RAG not only improves the quality of the answers provided by LLMs, but also significantly expands their knowledge base, allowing them to reference up-to-date and specific information that is not available in their initial training set [11]. GraphRAG is a potent enhancement to traditional RAG which is based on vector search [15]. This approach utilizes the organized structure of graph databases, where data is categorized into nodes and interconnected relationships, to enrich the quality of information retrieval [10]. By doing so, GraphRAG significantly enhances the depth and contextuality of the information gathered, providing a more nuanced and comprehensive understanding compared to standard methods [12,13].

Knowledge graphs, with their ability to represent entities (like players, teams, tournaments) and their relationships (team memberships, competition history) [2], offer a promising solution for efficiently retrieving information from sports datasets. This structured format allows for more nuanced queries and facilitates the retrieval of comprehensive information compared to traditional methods that rely on keyword matching in unstructured data [8]. Despite their potential, existing approaches for sports data retrieval might struggle with the sheer volume and complexity of sports data. Additionally, capturing the rich context and relationships within sports datasets can be challenging with traditional methods.

Our research aims to tackle the complexities of extracting information from extensive sports datasets using the power of knowledge graphs. We propose Soccer-GraphRAG, a framework designed for retrieving information from soccer knowledge graphs through natural language queries. Soccer-GraphRAG utilizes knowledge graphs created from textual soccer datasets, such as the recently released SoccerNet-Echoes dataset [5] which includes automated transcriptions of audio commentaries from soccer game broadcasts. By implementing an approach based on knowledge graphs, we enhance user interaction and improve the accessibility of sports datasets. The contributions of our work are the following:

- Soccer-GraphRAG is specifically designed to leverage Automatic Speech Recognition (ASR) data, which has traditionally been challenging due to its unstructured nature and the presence of transcription errors. Our approach enhances the accuracy and utility of information extracted from ASR sources by effectively integrating these data into our knowledge graph.
- Our research addresses the issue of hallucinations in ASR outputs where erroneous or misleading information might be generated. Through the structured organization of data in the knowledge graph and the contextual checks provided by the Soccer-GraphRAG framework, we minimize these inaccuracies.
- Our Soccer-GraphRAG framework is specifically designed for the soccer domain, enabling precise information retrieval from extensive sports datasets through natural language queries.

The rest of the article is organized as follows: Sect. 2 details our Soccer-GraphRAG method and strategy. Section 3 describes the experimental setup, including the source of the dataset, pre-processing data phase, and knowledge graph construction. Section 4 presents the preliminary results. Finally, Sect. 5 concludes the paper by summarizing the main outcomes and suggesting potential directions for future research.

2 Methodology

This section outlines the methodology of our framework for retrieving information from a knowledge graph in response to user queries. When a user asks a question about soccer, the system leverages an LLM to create an appropriate Cypher query. This Cypher query is designed to extract relevant data from the knowledge graph, which is constructed from our dataset to provide accurate and detailed information. Once the Cypher query is created, the smart search tool utilizes it to navigate through the knowledge graph and retrieve the relevant information. The retrieved information, along with the user's original question, is then fed back into the LLM. The LLM processes this input to generate a comprehensive response. This complete process, from query generation to the final response, is illustrated in Fig. 1.

The soccer knowledge graph illustrated in Fig. 1 is constructed through a process that begins with a pre-processing phase. In this initial stage, entities and their interconnections are identified and extracted from the data. These entities and relationships are then used to create nodes and edges of the soccer knowledge graph. This graph is then stored in the Neo4j graph database. For this research, we employed the Neo4j Python driver to establish and manage connections with the Neo4j database, enabling efficient query execution and manipulation of graph data. This integration is part of our Python-based analytical framework.

In summary, our system operates through the following steps:

- **Data Pre-processing:** Entities and relations are extracted from our dataset.
- **Graph Construction:** The pre-processed data is used to structure soccer knowledge graphs.
- **Query Translation:** User query is translated into Cypher queries.
- **Information Retrieval:** Cypher queries retrieve relevant soccer graph data from Neo4j.
- **Answer Generation:** Based on retrieved graph data, LLM generates a response.

This methodology integrates advanced LLMs and graph technology to create a dynamic and responsive system for answering user queries about soccer. By utilizing Cypher queries to interact with a well-structured knowledge graph, and

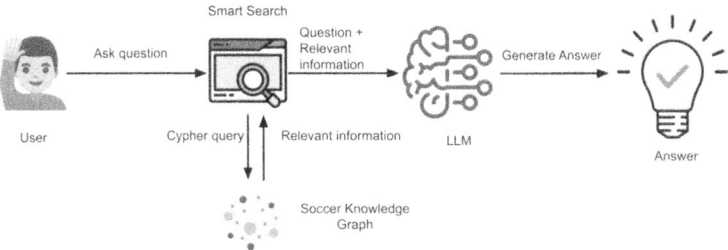

Fig. 1. Soccer-GraphRAG framework overview.

leveraging the processing power of LLMs for intelligent response generation, this framework ensures that users receive accurate, relevant, and detailed answers.

Listing 1.1. ASR data structure.

```
"segments": {
    "segment index": [
        "start time":  < >
        "end time":    < >
        "text":        < >
                       ],
    ...
        }
```

3 Experimental Setup

3.1 Dataset

SoccerNet [6] is an expansive dataset designed for the analysis and understanding of soccer videos, collected from 550 soccer games from 6 different European leagues and 4 different seasons, narrated live by commentators speaking one of 10 different languages. Gautam et al. [5] have extended this dataset with audio commentary transcriptions using ASR, and have released the transcriptions for each game half as an open dataset called SoccerNet-Echoes. Our data source is ASR dataset from SoccerNet-Echoes dataset. The data in ASR has the structure as described in Listing 1.1.

Figure 2 shows a sample of ASR data for an English Premier League (EPL) game played between Chelsea and Burnley on February 21st, 2015 [5]. As illustrated in Fig. 2 and Listing 1.1, each segment in an ASR file consists of a start timestamp, end timestamp, and corresponding text.

```
"segments": {
    "0": [
        "0.000",
        "3.000",
        "The duel has already started, Barley handles the ball."
    ],
    "1": [
        "3.000",
        "8.240",
        "It must be said that they also faced each other in the first
    ],
    "2": [
        "8.240",
        "9.680",
        "Chelsea won 1-3."
```

Fig. 2. Sample ASR file for an English Premier League (EPL) game played between Chelsea and Burnley on February 21st, 2015.

3.2 Data Pre-processing

The aim of the pre-processing phase is to extract 3-tuples from the 'text' in Listing 1.1, in the format of (entity1, relation, entity2). These 3-tuples are used for the construction of a graph in Sect. 3.3.

To extract 3-tuples, we create two different lists: one is used to extract entity1 and entity2, and the other is used for the relation. For entity1 and entity2, we create a list containing the names of players, teams, referees, and events in a soccer game, such as goals, yellow cards, corners, etc. To extract the relation elements, we create a list containing the different types of actions in a soccer game such as score, won, etc. Subsequently, we utilize these lists and employ GPT-3 to extract 3-tuples from 'text'. The extracted 3-tuples are collected in an array named 'entities'. Now, by adding the timestamp and game half, we create a new ASR dataset, which has the structure as shown in Listing 1.2. Figure 3 presents the corresponding 'new' ASR file for the original file shown in Fig. 2.

Listing 1.2. New ASR Data Structure.

```
"time": [
    "start time",
    "end time"
    ]
"description":  <"text">
"half": " "
"entities": {(entity1, relation, entity2)}
```

3.3 Graph Construction

For each game in our dataset, we create one graph. Below, we outline the detailed structure of these graphs.

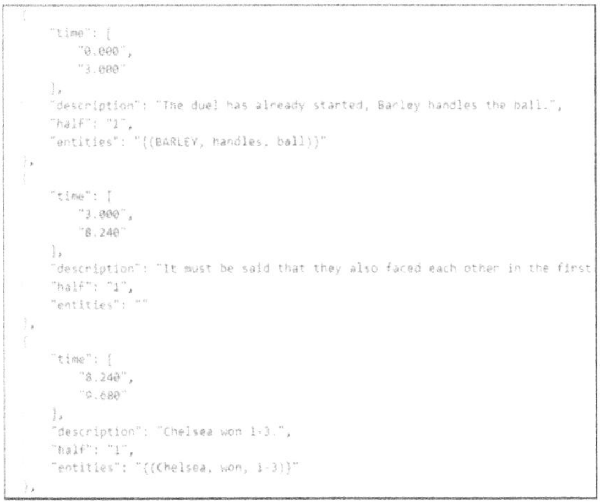

Fig. 3. Sample 'new' ASR file, updated from the original file depicted in Fig. 2.

Game Node & Team Node. For each game, we instantiate a node in Neo4j labelled Game with these attributes: away_team, home_team, awaya_score, home_score, coach_home_team, coach_away_team, date, venue, referee, winner. Each team is represented by a single node with label Team without any attributes.

Game & Team Nodes Connections. We establish connections between each team and the corresponding games using an edge labelled PARTICIPATED_IN. Additionally, teams are connected to their respective games with edges labelled HOME_TEAM or AWAY_TEAM, indicating the home or away status of the team. If the game result is not a draw, two additional edges, WINNER and LOSER, are created to link the game node with the winning and losing teams, respectively. Figure 4 demonstrates the connections between game and team nodes for an English Premier League (EPL) match between Crystal Palace and Arsenal on February 21st, 2015.

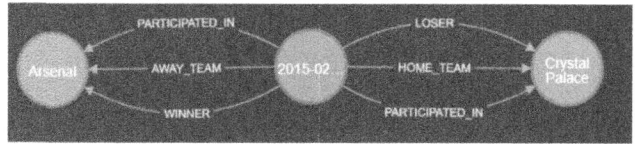

Fig. 4. Sample Game and Team nodes connections.

ASR Nodes. As discussed in the previous section, the new ASR dataset is utilized to construct the graph structure. To elucidate the methodology, we provide a detailed explanation using the example depicted in Fig. 5.

```
{
    "time": [
        "44.440",
        "46.800"
    ],
    "description": "Cuadrado who once again receives an opportunity as a starter.",
    "half": "1",
    "entities": "{(Cuadrado, receives_opportunity, starter)}"
},
```

Fig. 5. An example of ASR Data.

For the text in 'description', we create one node labelled Description, which includes the attributes time and half. Each element in 'entities' is represented as a triplet (entity1, relation, entity2). For each triplet, we generate three nodes: two nodes labelled Entity corresponding to entity1 and entity2, and one node labelled RELATION corresponding to the relation component.

Entity & Relation Nodes Connections. We connect the entity1 node to the relation node with an edge labelled RELATION, and similarly, connect the relation node to the entity2 node with another RELATION edge. In Fig. 6a, the connections in the triple (Cuadrado, receives_opportunity, starter) are shown.

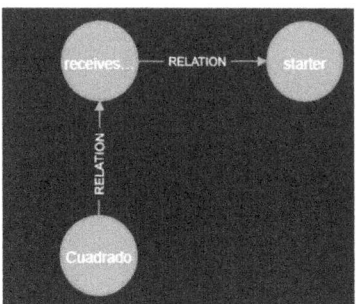 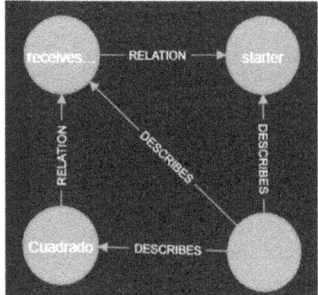

(a) Connections between Entity and Relation nodes.

(b) Connections between Entity, Relation, Description nodes.

Fig. 6. Visual representations of node connections.

The Description node is connected to these three nodes via an edge labelled DESCRIBES. Figure 6b illustrates this connection. Each Description, Relation, and Entity node is further connected to the corresponding game node via an edge labelled BELONGS_TO. Figure 7 demonstrates the connection between the text and the corresponding game, relation, and entity nodes.

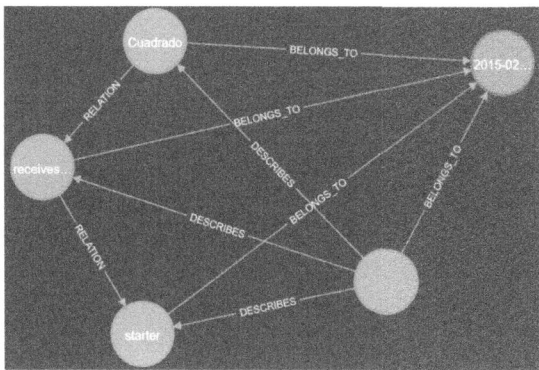

Fig. 7. An example of all connections.

```
> Entering new GraphCypherQAChain chain...
Generated Cypher:
MATCH (entity1:Entity)-[:RELATION]->(r:Relation)-[:RELATION]->(entity2:Entity)
WHERE r.name = 'scored_by'
RETURN entity1.name, entity2.name
Full Context:
[{'entity1.name': 'Felipe Luis', 'entity2.name': 'Chelsea'}, {'entity1.name': 'Felipe Luis', 'entity2.name': 'Ivanovic'}, {'e
ntity1.name': 'Felipe Luis', 'entity2.name': 'first'}, {'entity1.name': 'Felipe Luis', 'entity2.name': 'Felipe Luis'}, {'enti
ty1.name': 'Felipe Luis', 'entity2.name': "Ivanovic's fourth goal"}, {'entity1.name': 'Felipe Luis', 'entity2.name': 'goal'},
{'entity1.name': 'Ivanovic', 'entity2.name': 'Chelsea'}, {'entity1.name': 'Ivanovic', 'entity2.name': 'Ivanovic'}, {'entity1.
name': 'Ivanovic', 'entity2.name': 'first'}, {'entity1.name': 'Ivanovic', 'entity2.name': 'Felipe Luis'}]

> Finished chain.
 {'query': ' tell me who scored the goal',
  'result': 'Felipe Luis scored the goal.'}
```

Fig. 8. Sample Q&A application.

4 Preliminary Results

In this section, we demonstrate the applications of our soccer knowledge graph within the context of LLM. Our soccer knowledge graph facilitates the summarizing and analysis of teams' and players' performances within a single game, across an entire league, or throughout a season. Furthermore, this graph is adept at supporting question-answering tasks, showcasing its versatility and utility in processing complex sports data. ASR data are unstructured, and their direct utilization in question-answering systems, such as LLMs, can lead to inaccuracies and hallucinations. In our project, unstructured ASR data is converted into structured information through the Soccer-GraphRAG framework, leveraging a comprehensive knowledge graph. Our approach not only mitigates the typical inaccuracies found in direct analysis of ASR data but also significantly enhances the reliability and precision of the output. The following examples illustrate how our Soccer-GraphRAG framework leverages this enriched knowledge graph to deliver more accurate, reliable, and context-aware responses in a sports analytics setting.

Example 1: This example illustrates the application of our method for handling Q&A tasks. The query was, 'Tell me who scored the goal.' The LLM first converted this inquiry into a Cypher query, as depicted in Figs. 8 (Generated Cypher). This query was then used to interrogate the soccer knowledge graph, identifying pertinent nodes and edges. The execution details retrieved from the knowledge graph are detailed in the 'Full Context' section of Figs. 8. Utilizing this context, the LLM derived the answer, which is displayed in the 'Result' section of Figs. 8. Another example of Q&A task is shown in Fig. 9.

Example 2: This example illustrates the application of our method for handling summarization tasks. The query was, 'Tell me about Ivanovic and Cuadrado.' LLM first converted this inquiry into a Cypher query, as depicted in Figs. 10 (Generated Cypher). This query was then used to interrogate the soccer knowledge graph, identifying pertinent nodes and edges. The execution details retrieved from the knowledge graph are detailed in the 'Full Context' section of Figs. 10. Utilizing this context, the LLM derived the answer, which is displayed in the 'Result' section of Figs. 10.

```
MATCH (e1:Entity)-[:RELATION]->(r:Relation)-[:RELATION]->(e2:Entity)
WHERE r.name = 'injured_by'
RETURN e1.name, e2.name
Full Context:
[{'e1.name': 'OBI MIKEL', 'e2.name': 'OBI MIKEL'}]
> Finished chain.
{'query': 'tell me who injured in the game',
 'result': 'OBI MIKEL injured in the game.'}
```

Fig. 9. Sample Q&A application.

```
Generated Cypher:
MATCH (entity1:Entity)-[:RELATION]->(r:Relation)-[:RELATION]->(entity2:Entity), (d:Description)-[:DESCRIBES]->(r)
WHERE entity1.name = 'Ivanovic' AND entity2.name = 'Cuadrado'
RETURN r.name, d.name
Full Context:
[{'r.name': 'offers', 'd.name': 'Ivanovic offers Cuadrado in that unmarking to win the baseline.'}, {'r.name': 'unmarking_off
ers', 'd.name': 'Where Ivanovic is, Cuadrado offers himself in that unmarking to win the baseline.'}]

> Finished chain.
{'query': ' tell me about Ivanovic and Cuadrado',
 'result': 'Ivanovic offers Cuadrado in that unmarking to win the baseline. Cuadrado offers himself in that unmarking to win
 the baseline where Ivanovic is.'}
```

Fig. 10. Sample summarization application.

The examples provided demonstrate how the application of our methodology can lead to a more accurate and nuanced understanding of game dynamics and player performances. Furthermore, the use of a knowledge graph in LLM-based systems minimizes the risk of inaccuracies and hallucinations often associated with direct ASR data usage, thereby increasing confidence in the outputs generated by such systems. Looking forward, the refinement and expansion of our framework could significantly improve the efficiency of sports analytics, potentially influencing other fields where issues of unstructured data present similar challenges. The next phase of our project will focus on expanding the application of Soccer-GraphRAG to encompass larger datasets, including both structured and unstructured data, thereby testing the framework's versatility and scalability.

5 Conclusion

We present the Soccer-GraphRAG framework which aims to support the retrieval of soccer-related information via natural language queries. Relying on an experimental knowledge graph structure built upon the SoccerNet-Echoes dataset, the framework has shown promising performance in preliminary experiments. The next steps include the fine-tuning and up-scaling of the graph structure to reflect the entire dataset.

Acknowledgement. This work was partly funded by the Research Council of Norway, project number 346671 (AI-Storyteller), and has benefited from the Experimental Infrastructure for Exploration of Exascale Computing (eX3), which is financially supported by the Research Council of Norway under contract 270053.

References

1. Chang, Y., Wang, X., Wang, J., et al.: A survey on evaluation of large language models. ACM Trans. Intell. Syst. Technol. **15**(3), 1–45 (2024). https://doi.org/10.1145/3641289
2. Chen, Z., Zhang, Y., Fang, Y., et al.: Knowledge Graphs Meet Multi-Modal Learning: A Comprehensive Survey. arXiv (2024). https://doi.org/10.48550/arXiv.2402.05391
3. Gao, Y., Xiong, Y., Gao, X., et al.: Retrieval-Augmented Generation for Large Language Models: A Survey. arXiv (2023). https://doi.org/10.48550/arXiv.2312.10997
4. Gautam, S.: FactGenius: Combining Zero-Shot Prompting and Fuzzy Relation Mining to Improve Fact Verification with Knowledge Graphs. arXiv (2024). https://arxiv.org/abs/2406.01311
5. Gautam, S., et al.: SoccerNet-Echoes: A Soccer Game Audio Commentary Dataset. arXiv (2024). https://doi.org/10.48550/arXiv.2405.07354
6. Giancola, S., Amine, M., Dghaily, T., Ghanem, B.: SoccerNet: a scalable dataset for action spotting in soccer videos. In: 2018 IEEE/CVF Conference on Computer Vision and Pattern Recognition Workshops (CVPRW), pp. 18–22. IEEE (2018). https://doi.org/10.1109/CVPRW.2018.00223
7. Jeong, C.: A study on the implementation of generative AI services using an enterprise data-based LLM application architecture. advances in artificial intelligence and machine learning. Res. **3**(4), 1588–1618 (2023). https://oajaiml.com/uploads/archivepdf/43901191.pdf
8. Jia, R., Zhang, B., Méndez, S.J.R., et al.: Leveraging Large Language Models for Semantic Query Processing in a Scholarly Knowledge Graph. arXiv (2024). https://doi.org/10.48550/arXiv.2405.15374
9. Chen, J., Lin, H., Han, X., Sun, L.: Benchmarking large language models in retrieval-augmented generation. In: The Thirty-Eighth AAAI Conference on Artificial Intelligence (AAAI 2024) (2024). https://arxiv.org/pdf/2309.01431
10. Pan, J.Z., Vetere, G., Gomez-Perez, J.M., et al.: Exploiting Linked Data and Knowledge Graphs in Large Organisations. Springer, Cham (2017). https://doi.org/10.1007/978-3-319-45654-6
11. Siriwardhana, S., Weerasekera, R., Wen, E., et al.: Improving the domain adaptation of retrieval augmented generation (RAG) models for open domain question answering. Trans. Assoc. Comput. Linguist. **11**, 1–17 (2023). https://doi.org/10.1162/tacl_a_00530
12. Wei, L., Xinyan, X., Jiachen, L., Hua, W., Haifeng, W., Junping, D.: Leveraging graph to improve abstractive multi-document summarization. In: Proceedings of the 58th Annual Meeting of the Association for Computational Linguistics (2020). https://doi.org/10.18653/v1/2020.acl-main.555
13. Xu, W., Fang, M., Yang, L., et al.: Enabling language representation with knowledge graph and structured semantic information. In: International Conference on Computer Communication and Artificial Intelligence (CCAI). IEEE (2021). https://doi.org/10.1109/CCAI50917.2021.9447453
14. Yang, J., Jin, H., Tang, R., Han, X., Feng, Q., Jiang, H., et al.: Harnessing the power of llms in practice: a survey on ChatGPT and beyond. ACM Trans. Knowl. Discovery Data (2023). https://doi.org/10.1145/3649506
15. Ye, X., Yavuz, S., Hashimoto, K., et al.: RnG-KBQA: Generation Augmented Iterative Ranking for Knowledge Base Question Answering. arXiv (2021). https://doi.org/10.48550/arXiv.2109.08678

Enhanced Semantic Understanding with Graph-Based Information Retrieval

Giovanni M. De Filippis[1], Antonio M. Rinaldi[1], Cristiano Russo[1], and Cristian Tommasino[1,2(✉)]

[1] Department of Electrical Engineering and Information Technologies, University of Naples Federico II, Via Claudio, 21, 80125 Naples, Italy
{giovannimaria.defilippis,antoniomaria.rinaldi,cristiano.russo}@unina.it
[2] Interdepartmental Research Center on Management and Innovation in Healthcare (CIRMIS), University of Naples Federico II, Naples, Italy
cristian.tommasino@unina.it

Abstract. Traditional information retrieval systems, primarily based on keyword searches, often fail to capture nuanced domain-specific relationships. To address such an issue, we propose using a semantic graph-based approach, which enables enhanced semantic querying capabilities and contextual data retrieval. By mapping entities and their complex interdependencies within a graph, our system allows for sophisticated querying beyond simple keyword matches. It effectively leverages interconnected data to provide contextually relevant responses to complex queries, thereby improving the accuracy and depth of information retrieval. We have effectively demonstrated our solution in the field of nutrigenomics, highlighting how semantic graphs can significantly improve data interpretation and facilitate the creation of highly personalized nutrition plans and therapeutic interventions based on comprehensive and detailed genetic insights. Our code is publicly available on GitHub at https://github.com/CosmoIknosLab/bertopic_graph.

Keywords: Information Retrieval · Semantic Graphs · Semantic Search · Nutrigenomics

1 Introduction

In information retrieval, traditional lexical-based techniques are now being augmented by semantic methods that investigate the underlying meanings of the data. An auspicious aspect of this semantic expansion is the use of semantic graphs, which present knowledge in a structured form and enable a more profound comprehension of content relationships. These graphs amalgamate entities and their interrelations derived from the data, enabling more nuanced and context-aware retrieval systems. Our paper introduces an approach that employs a graph-based representation of information derived from topic modeling outputs, proving its effectiveness in nutrigenomics.

The core of a semantic graph lies in its ability to model complex relationships between various pieces of information through nodes (representing entities or concepts) and edges (representing relationships) [7]. This structure mirrors human cognitive processes and enhances computational algorithms to perform more efficiently in search, data mining, and information discovery tasks. In this context, such techniques have been successfully applied in various domains, such as medical imaging [17,18], cultural heritage [11,16], and semantic web [14,15].

The field of nutrigenomics, precisely personalized nutrition, offers new insights into how our genetic makeup influences dietary responses and predispositions to various health conditions [8]. The complexity of interpreting vast genomic datasets becomes apparent as we delve deeper into the genetic underpinnings of diet-induced conditions, such as oxidative stress and food sensitivities [9]. This complexity underscores the necessity for advanced computational tools to handle, analyze, and interpret these large datasets meaningfully. Advancements in natural language processing (NLP) and machine learning (ML) have opened new avenues for exploring these intricate genetic datasets. Deep learning models, such as the BERTopic [6] algorithm, have proven instrumental in distilling large-scale genomic and clinical data into actionable insights. These models leverage sophisticated algorithms to parse through extensive literature and clinical reports, identifying patterns and associations that might escape conventional analysis methods.

However, while topic modeling offers a structured overview of the vast landscape of genomic data, a critical need remains for enhanced semantic understanding and precise information retrieval systems. Traditional keyword-based searches often need to catch up when answering complex queries that require a deep understanding of context and relationships within the data. Addressing this pressing gap, our research proposes an innovative approach that utilizes a graph-based representation of information derived from topic modeling outputs to enable more nuanced and context-aware information retrieval. Therefore, our paper presents a novel semantic understanding and information retrieval system constructed from topics related to genetic polymorphisms and dietary responses. This system not only aids in identifying direct associations but also facilitates the exploration of complex queries involving multiple facets of nutrigenomics. By integrating additional NLP techniques, such as named entity recognition and relation extraction, our system remains dynamic and adaptive to new research findings, continually refining its understanding and enhancing its retrieval capabilities. This research has significant implications for the field of nutrigenomics, offering a more nuanced and context-aware approach to information retrieval.

We have structured the rest of the paper as follows: Sect. 2 provides an overview of the related work, setting the context for our research; Sect. 3 describes our proposed method in detail, outlining the steps we have taken to develop our innovative framework; Sect. 4 presents our experimental results on a nutrigenomics dataset, demonstrating the effectiveness of our approach; Sect. 5 offers a brief discussion on the implications of our findings and concludes, summarizing the key points of our research.

2 Related Work

This section briefly analyzes topic modeling and knowledge graphs to enhance information retrieval systems. The work of Whan [21] explores a knowledge-based information retrieval model incorporating a hierarchical thesaurus. This model calculates the conceptual distance between a query and an object, indexed with weighted terms from the hierarchical thesaurus. Wang et al. [20] proposed an innovative information retrieval method grounded in knowledge graphs. Their research underscores the limitations of keyword-based full-text searches in meeting users' search requirements, thereby advocating for applying knowledge graphs in information retrieval. Dietz et al. [3] examine the proliferation of publicly accessible and proprietary knowledge graphs (KGs), noting the increasing depth and breadth of content they offer. Their study emphasizes the potential of KGs in enhancing text-centric retrieval applications. Gaur et al. [5] introduced ISEEQ, a cutting-edge approach for generating Information Seeking Questions (ISQs) from succinct user queries within a large relevant text corpus. ISEEQ enriches user queries with knowledge graphs, retrieves pertinent context passages to formulate coherent ISQs adhering to a conceptual flow, and employs a novel deep generative-adversarial reinforcement learning-based technique for ISQ generation. Abu et al. [1] present a transferable framework designed to generate domain-specific explanations for intelligent information retrieval systems. This framework leverages knowledge graphs for domain knowledge modeling and incorporates graph-based components to generate textual and visual explanations of the retrieved information. Venkatesh et al. [19] propose a novel Multimodal Information Retrieval System utilizing a Knowledge Graph (MIR-KG) to address users' information needs. This approach processes multimodal input, employs an offline knowledge graph representation called ontology to identify entities and relations, and generates dynamic KG queries. A comprehensive survey by Reinanda et al. [13] discusses the multifaceted benefits that modern Information Retrieval (IR) systems can derive from the information contained in Knowledge Graphs (KGs), regardless of whether these KGs are publicly available or proprietary. Our framework harnesses the capabilities of knowledge graphs to augment information retrieval, thereby addressing critical limitations inherent in traditional keyword-based search methodologies. As evidenced by recent studies such as those by Wang et al. [20] and extensively reviewed by Reinanda et al. [13], the extensive content within KGs offers a robust structure for capturing intricate interrelations between entities and concepts. This structure significantly enhances retrieval accuracy by aligning with the contextual and semantic complexities of user queries. Our approach is further refined by incorporating multimodal inputs and generating dynamic, domain-specific KG queries, as explored by Venkatesh et al. [19], enabling our system to meet traditional information needs and address sophisticated queries involving complex interactions and detailed explanations, as highlighted by Abu et al. [1].

By adopting a knowledge graph-based framework, we aim to surpass the limitations of existing models, offering a more intuitive and context-aware retrieval experience that meets users' evolving expectations in the digital age.

3 The Proposed Framework

The framework implemented to build a semantic graph, depicted in Fig. 1, significantly enhances information retrieval and semantic understanding.

We implemented a data analysis pipeline to investigate themes in a repository of scientific papers. We selected the dataset based on its extensive coverage of relevant content and its alignment with our research objectives. The code implemented for this study is openly available on GitHub[1].

3.1 Data Source and Preprocessing

Our methodology begins by identifying and categorizing thematic content from comprehensive dataset literature into discernible topics. Subsequently, these topics serve as the nodes within the semantic graph. We focused our case study on the field of nutrigenomics, taking as a data source the GRPM dataset [2], a dataset based on this literature domain enriched of relative genetic features [2]. We refined a subset of the GRPM dataset for our use case by exploiting its labeling based on MeSH (Medical Subject Headings) ontology. A domain expert in the field selected these ontology terms representing specific areas within nutrigenomics, such as food intolerances, allergies, diet-induced oxidative stress, and xenobiotic metabolism. Subsequently, the textual data extracted (37,042 paper abstracts) was prepared for topic modeling.

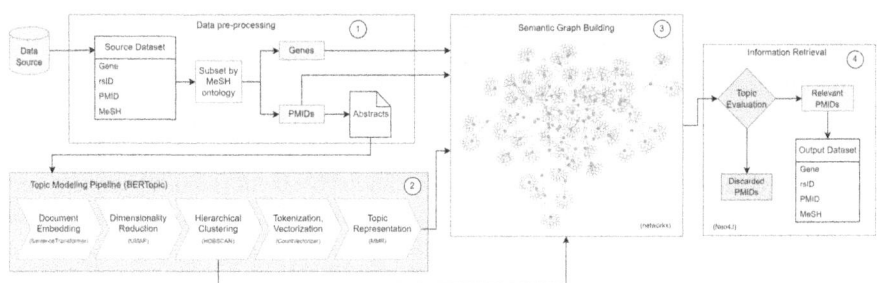

Fig. 1. Proposed Workflow

3.2 Topic Identification and Modeling

The core of our framework is topic identification, which is required to build the KG. The transformation of textual data into numerical embeddings is based on BERTopic [6]. Studies have shown that BERTopic offers more specificity and stability than LDA, making it a versatile tool for topic analysis [4]. Subsequently, we

---

[1] https://github.com/CosmoIknosLab/bertopic_graph.

applied UMAP [10], a dimensionality reduction technique, to transform the document embeddings into a more manageable form for the clustering algorithm, aiming to generate high-quality clusters. Then, we employed the HDBSCAN clustering technique, a sophisticated algorithm that performs density-based spatial clustering, adapting to clusters with different densities. This strategy enhances cluster differentiation while reducing noise, making it particularly adept at detecting outlier clusters. Additionally, it automates the process of determining the number of clusters, presenting substantial advancements compared to traditional clustering methods. The topic modeling phase involved transforming the clustered text data into a numerical matrix, where each document representation comprised features reflecting word frequency. We computed term frequency-inverse document frequency (c-TF-IDF) that accounts for term relevance inside each document cluster, allowing us to pinpoint and quantify keywords within each topic.

Creating the semantic graph involves establishing links between nodes via edges representing contextual relationships and dependencies among the identified topics. This sophisticated graph-based structure enables advanced querying capabilities during the information retrieval phase.

3.3 Building the Semantic Graph

Following the topic modeling, we constructed a heterogeneous semantic graph based on the clustered topics. The graph contains three types of nodes that represent terms, topics, and documents. Terms are linked with the relative topic, but documents are linked to the associated topic.

We employed hierarchical clustering of the topics to define the inter-topic relationships within the semantic graph. Specifically, we built a Tree-Based Hierarchical Graph (TBHG) to model multiple documents, similar to the approach proposed by Zheng et al. [22].

A TBHG integrates two structures: a tree-based hierarchy H and a graph G at each level of this tree. The tree structure represents the parent-child (or ascendant-descendant) relationships among nodes at different levels, whereas the graphs represent peer relationships among nodes at the same level.

The construction of a TBHG involves two main steps: first, formulating a tree-based hierarchy (H), and second, building graphs (G) at each level.

We used a density-based clustering algorithm to hierarchically cluster the document embeddings into several topic groups (a set of clusters C) in a divisive manner. Consequently, similar documents are grouped into the same clusters across different levels.

A tree-based hierarchy, denoted by H, consists of clusters organized into different levels specified by L (a set of levels) and C (the collection of clusters). Each cluster belongs to a specific level within this hierarchy. A TBHG extends this concept by integrating the hierarchy H with an additional graph structure G. This graph G consists of vertices and edges corresponding to the clusters and their connections at each level of the hierarchy.

This graphical representation visually captures the relationships and hierarchies among the topics, capturing the complex interactions and thematic structures within the dataset. The semantic graph's construction enabled a structured exploration of the data, providing insights into the interconnectedness of topics and the broader thematic landscape.

Utilizing the semantic graph, we implemented an information retrieval system to extract and synthesize information relevant to specific research inquiries. This phase involved querying the semantic graph to identify and retrieve topic clusters and associated documents that are most pertinent to the genetic factors influencing dietary adverse reactions. The retrieval process was optimized to leverage the structural and semantic insights provided by the graph, enabling precise and contextually relevant information extraction. For more details, please refer to Sect. 4.2.

3.4 Advanced Visualization and Topic Refinement

Our analysis also included advanced visualization tools to facilitate an interactive exploration of the topics and their relationships. These visualizations helped illustrate the hierarchical structure and interconnections between different topics, providing deeper insights into the dataset's thematic architecture. In the final stages of our study, we focused on refining the topic representations using a generative model, GPT-4, to improve their semantics. Such a model utilized the prominent terms within each topic to generate descriptive and concise titles. This step was crucial in interpreting the topics relevant to our research objectives and selecting specific genetic variations relevant to personalized nutrition interventions.

3.5 Linking Topics with Genetic Data

Moreover, we linked the identified topics with specific cited genes and relevant literature references inside the graph itself, establishing a direct correlation between our thematic findings and the underlying genetic data. This comprehensive approach not only enhanced the clarity of our results but also underscored their applicability in the context of dietary adverse reactions influenced by genetic factors [12].

This semantic graph approach facilitates the effective navigation of intricate genomic data and optimizes the retrieval process, ensuring that users can access and interpret the vast information more efficiently. By employing topic modeling as a preparatory step, we enhance the semantic graph's utility, making it a potent tool for researchers to derive actionable insights and make informed decisions in nutrigenomics. This methodological framework underscores the potential of semantic graphs in transforming the landscape of information retrieval, providing a deeper understanding of the data within scientific research.

4 Experimental Results

This section presents the experimental results of the implementation of the semantic graph-based information retrieval system based on the BERTopic algorithm in the nutrigenomics domain. We provide statistics on the source dataset used in our analysis, followed by the outcomes of the topic modeling process and retrieval based on semantic graph querying (Table 1).

Table 1. Output Dataset Statistics

	Gene	rsID	PMID	MeSH
Source Dataset	15,519	300,239	260,587	21,755
Filtered by MeSH	8,280	45,456	37,042	251
Graph-Based Information Retrieval	3.961	14.555	10.221	199

4.1 Hierarchical Clustering Tree

We used the hierarchical clustering tree (Fig. 2) as the backbone of the semantic graph by aggregating semantically associated topics into coherent clusters.

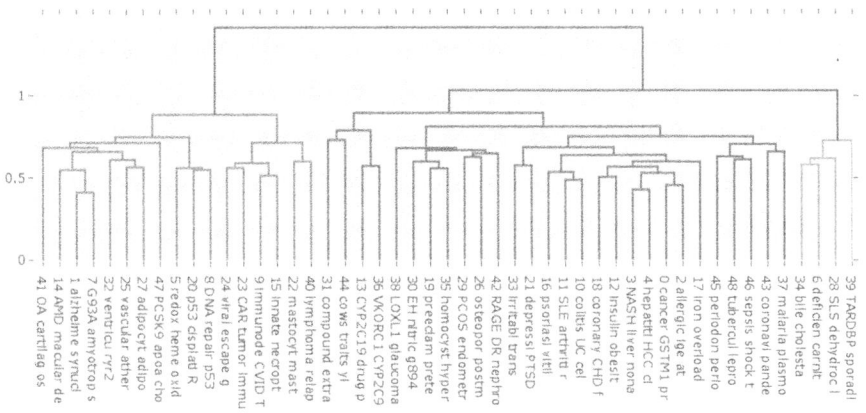

Fig. 2. Hierarchical Clustering Tree

The visualization offered illustrates the arrangement of identified topics into specific clusters, highlighting their semantic connections and thematic unity.

4.2 Semantic Graph for Information Retrieval

The semantic graph serves as the foundation of the information retrieval system, encapsulating the intricate relationships and dependencies between topics. By querying the semantic graph, researchers can navigate through the interconnected topics and retrieve relevant information with contextual understanding. In our use case, we retrieved 21 relevant topics to which meaningful names were assigned through the OpenAI GPT-4 model, considering the top keywords from the c-TF-IDF values of each topic (see Table 2).

In Fig. 3, a partial representation of the semantic graph constructed is depicted. The green numerical nodes represent the topics (see Table 2). The grey nodes are higher-level hierarchy nodes. Only terms for the most relevant topics to the nutrigenetic domain are depicted and the thickness of the edges is proportional to the cTF-IDF score for each term. It can be observed that some terms are common across multiple topics.

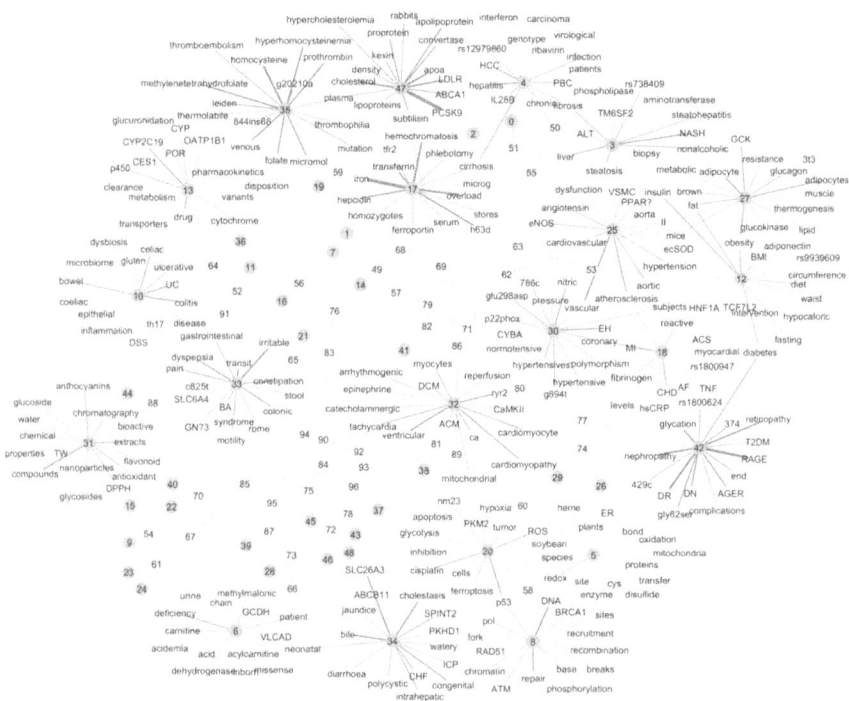

Fig. 3. Semantic Graph Extract (Color figure online)

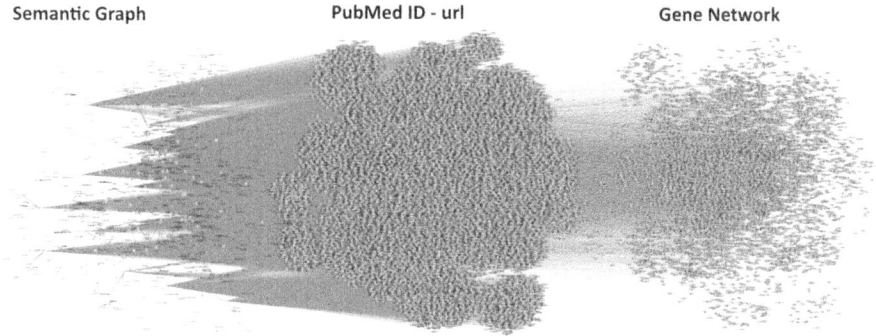

Fig. 4. Semantic Graph Layers

The complete graph consists of three layers (Fig. 4). The second layer of the graph contains nodes related to the articles with PubMed URLs, while the third layer comprises the complex network of genes associated with each cluster of articles. The complete graph can be downloaded from the GitHub repository "data" section and loaded in tools like Neo4j graph database for querying or Cytoscape to visualize and analyze the graph data interactively.

4.3 Topic Distribution Visualizations and Analysis

The visualizations and analyses comprehensively understand the topic distribution within the retrieved documents.

Figure 5 illustrates the projection of a subset of retrieved documents (5%) into a two-dimensional space, providing a condensed representation of the retrieved documents.

Figure 6 presents the similarity matrix that quantifies the semantic relationships and coherence between the selected topics. This heatmap aids in identifying topic clusters with higher inter-topic relevance and connectivity, offering insights into the pairwise similarities among the topics.

To provide a temporal perspective, Fig. 7 showcases the temporal distribution and evolution of topic occurrences within the retrieved topics. Analyzing the trend of topic frequencies over time can reveal emerging themes, shifts in research focus, and the relevance of topics across different time periods.

The combination of these visualizations and analyses offers a comprehensive overview of the performance of the semantic graph-based information retrieval system and its ability to facilitate nuanced semantic understanding and context-aware data retrieval in nutrigenomics research.

Table 2. Legend of the most representative topics, including the number of associated PubMed IDs (PMIDs).

Topic	PMIDs	Custom Name GPT-4
3	1157	Genetic and Metabolic Factors in Liver Diseases
4	1076	Hepatitis C: Fibrosis, Genotypes and Treatment Outcomes
5	1049	Redox Reactions and Enzyme Functions in Mitochondria
6	1007	Inborn Metabolic Disorders: Acidemia and Enzyme Deficiencies
8	887	DNA Repair Mechanisms and Chromatin Dynamics
10	740	Inflammatory Bowel Disease: Microbiome and Immune Response
12	665	Adiponectin, Obesity, and Diabetes: Genetic Associations
13	597	Drug Metabolism: Cytochrome P450 and Pharmacokinetic Variants
17	393	Iron Overload and Hemochromatosis: Genetics and Management
18	367	Inflammatory Markers and Coronary Heart Disease Risk
20	358	Cellular Redox Balance, Ferroptosis, and Tumor Progression
25	300	PPARγ Role in Vascular Function and Hypertension
27	261	Obesity, Adipocytes, and Metabolic Regulation Mechanisms
30	197	Genetic Polymorphisms in Hypertension and Vascular Function
31	192	Bioactive Compounds: Extraction, Antioxidants, and Nanoparticles
32	192	Cardiomyopathy and Arrhythmias: Cellular and Molecular Mechanisms
33	185	Irritable Bowel Syndrome: Pathophysiology and Genetic Insights
34	179	Cholestasis Syndromes: Genetics and Clinical Spectrum
35	178	Hyperhomocysteinemia and Thrombophilia: Genetic Associations
42	127	Diabetes Complications: RAGE and Glycation End Products
47	114	PCSK9 and Cholesterol Regulation: Impacts on Hypercholesterolemia

5 Discussion

In this study, we employed advanced topic modeling techniques in genomic literature to investigate the intricate relationships between genetic determinants and dietary responses. Our investigation demonstrated the effectiveness of using textual vector embeddings to query a semantic graph for information retrieval. Leveraging the semantic graph-structured representation of complex relationships and topic dependencies, we enhanced the accuracy and depth of information retrieval in the field of nutrigenomics.

The gene network embedded within the graph (Fig. 4) provides a valuable reference for personalized nutrition plans and tailored dietary recommendations by identifying genetic markers associated with nutrient metabolism. By integrating genetic data from semantic graph-based searches with nutritional guidelines, healthcare providers can design customized nutrition plans to optimize nutrient absorption, address dietary sensitivities, and mitigate disease risks [12].

The topic modeling and selection process yielded 21 key topics within nutrigenomics (see Table 2). High PubMed ID counts in topics like "Genetic and

Enhanced Semantic Understanding with Graph-Based IR   21

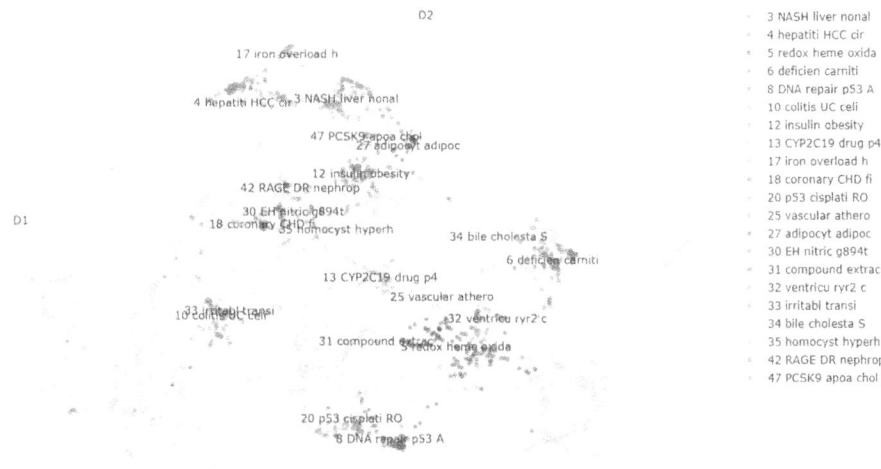

Fig. 5. Two-Dimensional Document Mapping

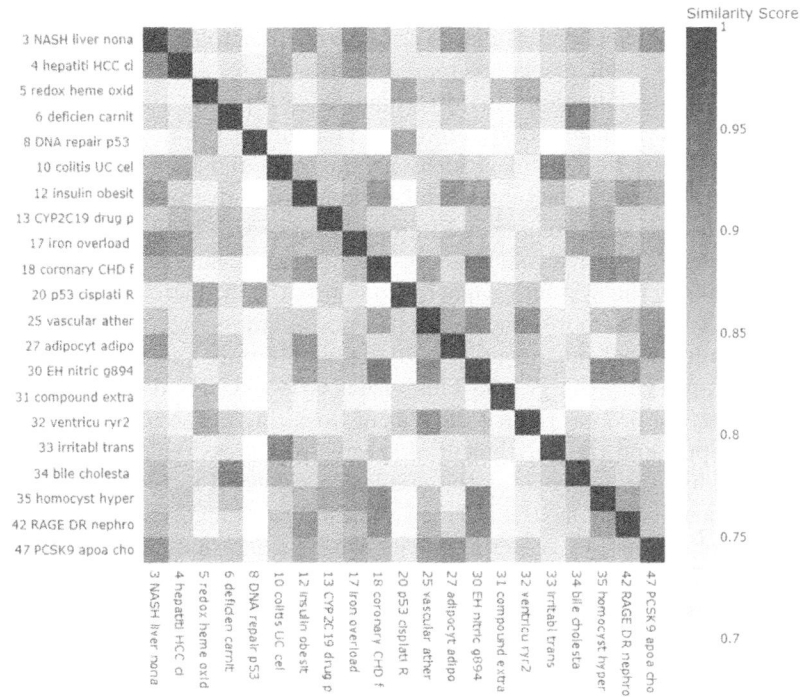

Fig. 6. Matrix of Similarity Indices

Fig. 7. Frequency of Topics Over Time

Metabolic Factors in Liver Diseases" (Topic 3, 1157 PMIDs) and "Hepatitis C: Fibrosis, Genotypes, and Treatment Outcomes" (Topic 4, 1076 PMIDs) indicate substantial research focus, providing useful indications about metabolic and genetic factors in liver-related diseases. Moreover, complex biochemical pathways and disease mechanisms are represented in topics such as "Redox Reactions and Enzyme Functions in Mitochondria" (Topic 5, 1049 PMIDs) and "DNA Repair Mechanisms and Chromatin Dynamics" (Topic 8, 887 PMIDs), which offer insights into cellular functions and potential therapeutic interventions. At last, topics on metabolic and inflammatory diseases like "Inflammatory Bowel Disease: Microbiome and Immune Response" (Topic 10, 740 PMIDs) and "Adiponectin, Obesity, and Diabetes: Genetic Associations" (Topic 12, 665 PMIDs) highlight the system's capacity to address multi-factorial health issues and potentially lead to more effective treatments and improved patient outcomes.

A two-dimensional document (Fig. 5) highlights only the topics that were retrieved by querying the graph. This visualization clearly demonstrates how the topic selection process refined the output, resulting in the pruning of certain areas (grey dots) and the enrichment of specific plot areas (colored dots).

Additionally, the analysis of the similarity values (Fig. 6) emphasizes the semantic coherence among the selected topics. This semantic coherence offers practical applications; by selecting topics with higher similarity indices, it is possible to retrieve the subnet of genes and polymorphisms associated with these topics. This can guide researchers in developing custom gene panels for genotype-based dietary recommendations [12].

Finally, the analysis of literature trends highlights the shifting focus in research areas, with a noticeable increase in interest in nonalcoholic steatohepatitis (NASH), mitochondrial function, and inflammatory bowel disease (IBD) up to 2020. This trend analysis underscores the dynamic nature of research priorities and the importance of continual examination to stay updated on emerging topics and developments in the field, as shown in Fig. 7.

6 Conclusion

Our study exemplified the effectiveness of integrating semantic technologies and text encoders, such as BERT, to extract valuable insights from complex genetic literature. A potential limitation of the method could be attributed to the noise removal executed by the HDBSCAN algorithm. The grouping of unclassified outliers could potentially result in the loss of some significant information. Acknowledging and addressing this limitation offers opportunities for future research to refine outlier analysis methods, striking a balance between comprehensive data analysis and noise reduction.

In conclusion, our research contributes to advancing biomedical information retrieval by harnessing semantic technology to decode the intricate interplay between genetic features and nutritional responses, as demonstrated in our nutrigenomics use case. This innovative approach enhances the understanding of genetic influences on dietary outcomes and paves the way for further advancements in personalized nutrition interventions and genomic data interpretation. Due to the promising results obtained in the nutrigenomics use case in future work, we will generalize our approach to other domains to prove its effectiveness.

Acknowledgments. We acknowledge financial support from the PNRR MUR project PE0000013-FAIR.

Disclosure of Interests. The authors declare that they have no competing interests.

References

1. Abu-Rasheed, H., Weber, C., Zenkert, J., Dornhöfer, M., Fathi, M.: Transferrable framework based on knowledge graphs for generating explainable results in domain-specific, intelligent information retrieval. In: Informatics, p. 6. MDPI (2022)
2. De Filippis, G.M., Monticelli, M., Pollice, A., Angrisano, T., Hay Mele, B.H., Calabro, V.: GRPM Dataset (2023). https://doi.org/10.5281/zenodo.8205724
3. Dietz, L., Kotov, A., Meij, E.: Utilizing knowledge graphs for text-centric information retrieval. In: The 41st International ACM SIGIR Conference on Research & Development in Information Retrieval, pp. 1387–1390 (2018)
4. Egger, R., Yu, J.: A topic modeling comparison between LDA, NMF, Top2Vec, and BERTopic to demystify twitter posts. Front. Sociol. **7**, 886498 (2022). https://doi.org/10.3389/fsoc.2022.886498
5. Gaur, M., Gunaratna, K., Srinivasan, V., Jin, H.: ISEEQ: information seeking question generation using dynamic meta-information retrieval and knowledge graphs. In: Proceedings of the AAAI Conference on Artificial Intelligence, pp. 10672–10680 (2022)
6. Grootendorst, M.: BERTopic: Neural topic modeling with a class-based TF-IDF procedure (2022). https://doi.org/10.48550/arXiv.2203.05794. arXiv:2203.05794
7. Ji, S., Pan, S., Cambria, E., Marttinen, P., Philip, S.Y.: A survey on knowledge graphs: representation, acquisition, and applications. IEEE Trans. Neural Netw. Learn. Syst. **33**(2), 494–514 (2021)

8. Kiani, A.K., et al.: Polymorphisms, diet and nutrigenomics. J. Prev. Med. Hyg. **63**(2 Suppl. 3), E125–E141 (2022). https://doi.org/10.15167/2421-4248/jpmh2022.63.2S3.2754
9. Mathers, J.C.: Nutrigenomics in the modern era. Proc. Nutr. Soc. **76**(3), 265–275 (2017). https://doi.org/10.1017/S002966511600080X
10. McInnes, L., Healy, J., Melville, J.: UMAP: Uniform Manifold Approximation and Projection for Dimension Reduction (2020). https://doi.org/10.48550/arXiv.1802.03426. arXiv:1802.03426
11. Ranjgar, B., Sadeghi-Niaraki, A., Shakeri, M., Rahimi, F., Choi, S.M.: Cultural heritage information retrieval: past, present and future trends. IEEE Access (2024)
12. Reddy, V.S., Palika, R., Ismail, A., Pullakhandam, R., Reddy, G.B.: Nutrigenomics: opportunities & challenges for public health nutrition. Indian J. Med. Res. **148**(5), 632–641 (2018). https://doi.org/10.4103/ijmr.IJMR_1738_18
13. Reinanda, R., Meij, E., de Rijke, M., et al.: Knowledge graphs: an information retrieval perspective. Found. Trends® Inf. Retrieval **14**(4), 289–444 (2020)
14. Rinaldi, A.M., Russo, C., Tommasino, C.: Visual query posing in multimedia web document retrieval. In: 2021 IEEE 15th International Conference on Semantic Computing (ICSC), pp. 415–420. IEEE (2021)
15. Rinaldi, A.M., Russo, C., Tommasino, C.: Web document categorization using knowledge graph and semantic textual topic detection. In: Gervasi, O., et al. (eds.) ICCSA 2021. LNCS, vol. 12951, pp. 40–51. Springer, Cham (2021). https://doi.org/10.1007/978-3-030-86970-0_4
16. Rinaldi, A.M., Russo, C., Tommasino, C.: An approach based on linked open data and augmented reality for cultural heritage content-based information retrieval. In: Gervasi, O., Murgante, B., Hendrix, E.M.T., Taniar, D., Apduhan, B.O. (eds.) ICCSA 2022. LNCS, vol. 13376, pp. 99–112. Springer, Cham (2022). https://doi.org/10.1007/978-3-031-10450-3_8
17. Rinaldi, A.M., Russo, C., Tommasino, C.: Effects of color stain normalization in histopathology image retrieval using deep learning. In: 2022 IEEE International Symposium on Multimedia (ISM), pp. 26–33. IEEE (2022)
18. Tommasino, C., Merolla, F., Russo, C., Staibano, S., Rinaldi, A.M.: Histopathological image deep feature representation for CBIR in smart PACS. J. Digit. Imaging **36**(5), 2194–2209 (2023)
19. Venkatesh, P.R., Chaitanya, K., Kumar, R., Krishna, P.R.: Conversational information retrieval using knowledge graphs. In: CIKM Workshops (2022)
20. Wang, C., Yu, H., Wan, F.: Information retrieval technology based on knowledge graph. In: 2018 3rd International Conference on Advances in Materials, Mechatronics and Civil Engineering (ICAMMCE 2018), pp. 291–296. Atlantis Press (2018)
21. Whan Kim, Y., Kim, J.H.: A model of knowledge based information retrieval with hierarchical concept graph. J. Doc. **46**(2), 113–136 (1990)
22. Zheng, Y., Zhang, L., Xie, X., Ma, W.Y.: Mining interesting locations and travel sequences from GPS trajectories. In: Proceedings of the 18th International Conference on World Wide Web, WWW 2009, pp. 791–800. Association for Computing Machinery, New York (2009). https://doi.org/10.1145/1526709.1526816

Identifying Shopping Intent in Product QA for Proactive Recommendations

Besnik Fetahu[✉], Nachshon Cohen, Elad Haramaty, Liane Lewin-Eytan, Oleg Rokhlenko, and Shervin Malmasi

Amazon.com, Inc., Seattle, WA, USA
besnikf@amazon.com

Abstract. Voice assistants have become ubiquitous in smart devices allowing users to instantly access information via voice questions. While extensive research has been conducted in question answering for voice search, little attention has been paid on how to enable proactive recommendations from a voice assistant to its users. This is a highly challenging problem that often leads to user friction, mainly due to recommendations provided to the users at the wrong time. We focus on the domain of e-commerce, namely in identifying *Shopping Product Questions* (SPQs), where the user asking a product-related question may have an underlying shopping need. Identifying a user's shopping need allows voice assistants to enhance shopping experience by determining *when* to provide recommendations, such as *product* or *deal* recommendations, or *proactive shopping actions recommendation*. Identifying SPQs is a challenging problem and cannot be done from question text alone, and thus requires to infer latent user behavior patterns inferred from user's past shopping history. We propose features that capture the user's latent shopping behavior from their purchase history, and combine them using a novel Mixture-of-Experts (MoE) model. Our evaluation shows that the proposed approach is able to identify SPQs with a high score of F1 = 0.91. Furthermore, based on an *online* evaluation with real voice assistant users, we identify SPQs in real-time and recommend shopping actions to users to add the queried product into their shopping list. We demonstrate that we are able to accurately identify SPQs, as indicated by the significantly higher rate of added products to users' shopping lists when being prompted after SPQs vs random PQs.

Keywords: Intent Detection · Shopping Intent Detection on Voice Queries · Customer Behavior Understanding

1 Introduction

Voice assistants, like Alexa or Google Assistant provide ubiquitous services through a variety of devices (e.g. smart speakers, phones, TVs etc.). Users interact with them for different purposes [34,43] such as question answering [11], task completion [9], conversational search [44], entertainment, or control of smart devices. Determining the underlying *user intent* is an active field of research [27,48,54], given that new skills are continuously added to voice assistants.

In this context, making proactive follow-on suggestions is an active area of research [12]. However, key challenges with voice-based conversational recommender systems remain, such as their failure to adapt to evolving user behavior [38]. Additionally, interactions are typically initiated by users, and proactive system recommendations may lead to increased user friction[1]. Knowing *when* to proactively recommend actions or items to users on the next turn, such as suggesting the right product recommendations [8,19,29,32,35,37] or next actions, is tightly dependent on accurately identifying the user's underlying intent. Correctly identifying this intent can avoid user dissatisfaction by providing recommendations only when necessary. Our work focuses on identifying the right time to make a proactive personalized recommendation in a voice-based conversational system.

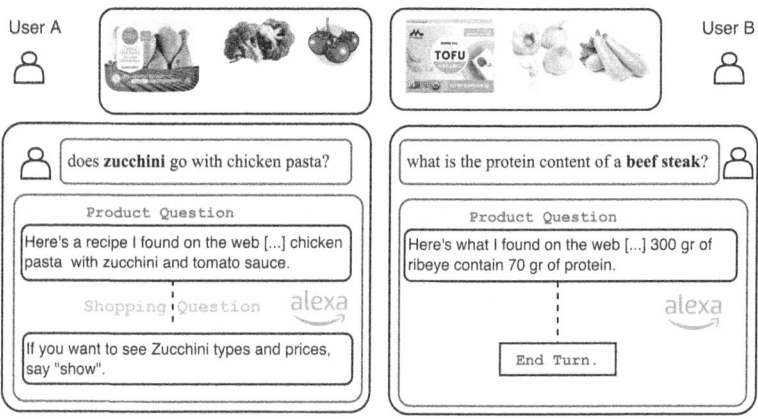

Fig. 1. User A searches for ingredients for a *recipe*. Our approach identifies the user having a shopping need, hence, triggering voice assistant's shopping recommendations. For User B's (likely vegetarian) our model predicts no shopping intent, hence, after answering, the turn is over.

E-commerce is an important functionality in voice assistants. In some contexts, the shopping intent can be explicit, such as in product searches (e.g. "*Buy me dog food*", or important functionality "*Search for an Apple watch stand*"). However, intent cannot always be captured from the question alone, e.g., product-related questions, such as: "*What can I use as a substitute for toilet paper?*"; "*How long do you cook rice?*;" *Are dogs allowed to eat tuna?*.

---

[1] User friction refers to the event whereby introducing a proactive experience we hinder the user in their task instead of helping.

The examples show that the user intent is often not clear from the Product Question (PQ) itself. PQs may emerge from an underlying desire to potentially purchase a product (*shopping need*), or only for a *general* knowledge need.

Voice product search [5,41] and question answering [6,16,39,40] have received attention, while understanding the reasons why users inquire information about certain products, especially in terms of *shopping need*, has seen very little progress. This is the main challenge that we tackle, namely determine whether a user has a shopping need. This allows voice assistants to determine *when* to surface recommendations that improve user's shopping experience, e.g. after answering a product question, show product recommendations, deals or promo suggestions, or offering additional product details for further examination. In this work, we do not focus on *what* to recommend given the vast literature in product (sequential) recommendation [8,19,29,32,35,37].

Figure 1 shows a hypothetical user interaction with a voice assistant. Depending on the user's interests, shopping recommendations may differ. For instance, for *User A*, based on their purchase history we can infer that they may be interested in purchasing a queried product, recommending shopping related actions, and thus ease the shopping process for the user. For *User B* on the other hand, who has vegetarian preferences, the intent is general knowledge need.

Identifying Shopping Product Questions (SPQ) in voice search poses several challenges. Contrary to shopping intent prediction in e-commerce sites [10,15,31], signals such as click-rate, browse time, hardware related gestures are not present. Furthermore, shopping need detection in voice search is inherently harder given the multi-purpose use of voice assistants, compared to a restricted use of e-commerce sites.

We propose an approach based on Graph Attention Networks (GAT) [50] to identify shopping need. Questions are considered nodes and are connected if they share the same product. As input features, we consider features that are geared at capturing diverse and latent aspects of shopping need, such as *product information*, user *past purchasing behavior*, and the *question* itself. Finally, based on the mixture-of-experts approach (MoE) [45], we propose a mechanism to compute a joint node representation from the diverse features.

Experimental evaluation on more than 370k voice assistant user questions shows that our proposed approach allows us to achieve highly accurate results in distinguishing SPQs. Furthermore, we carry out an online experiment with real voice assistant users, where we identify SPQs in real-time, and recommend users to add the queried product into their *shopping lists* following a SPQ, identified with high accuracy.[2] The approach presented in this paper through detection of SPQs allows voice assistants to recommend personalized responses according to user's interests and needs, such as *"check price"*, *"show deals"*, *"add to cart"*, *"compare products"* etc. Our main contributions in this work are as follows:

- A new problem definition for identifying SPQs that enable voice assistants to determine when to recommend to users.

---

[2] As future work we plan to experiment with different types of recommended shopping actions and assess the impact on the user experience.

– An approach to identify SPQs in voice search, using a novel way to combine diverse features through Mixture of Experts stemming from various user signals and containing different feature types.
– As part of these experiments, we perform detailed ablations showing the impact of the different features in identifying queries with shopping need, highlighting the latent nature of users shopping behavior, and thus opening directions for improving users shopping experience with voice assistants.
– Detailed offline and online experimental evaluation on voice assistant users and user queries.

2 Related Work

Users Needs in E-commerce. Informational need in e-commerce and voice shopping is typically expressed by asking questions about products. This domain is also referred to as Product Question Answering (PQA), and has attracted much interest in the past years [6,7,14,24,42,56]. Some research in the PQA domain has also been conducted with respect to the voice medium [6,16,39,40], where queries have different characteristics, and where cross-lingual and speech-to-text components add another layer of complexity. Most of the works in the PQA domain focus on different answering approaches and on different types of questions [25,26,36,42,53,55]. Here, we do not focus on the answering side at all, but rather focus on the informational need behind product questions.

Transactional need in e-commerce providers has been investigated, mainly through purchase intent prediction [2,10,15,22,31,46], where behavior features correspond to browse or hardware related (touch interaction on smartphones) [15]. Esmeli et al. [10] extract features such as product click rate, view time, number of visits, to predict purchase intent. Our key differences with these works is that we tackle the problem of voice modality, where voice assistants are used for highly diverse purposes [34,43], which is not the case of dedicated e-commerce sites, where the primary purpose is for shopping. Secondly, features that can be extracted from Web sites are not available for voice assistants, making the problem of predicting shopping needs of voice users more challenging. The work of [18] explores the problem of predicting the purchase rate of a ranking model in a voice setting, where there is an extreme position bias towards the first offer. This work lies in the domain of voice product search, where the user's intent is clearly transactional – unlike our work, which aims to identify an underlying shopping need in the voice PQA domain.

Product Search and Recommendations. There has been a large body of research in product search and product recommendations [3,28,30,51,57,58], which are crucial in the shopping domain, for providing relevant products, and recommendations reflecting users' preferences. The voice medium has also attracted research in this context [5,41], as shopping queries and behavioral shopping patterns in voice are different than in web [20]. Learning product embeddings is another challenge that has received much attention [17,47,49], as having relevant product embeddings allows for more accurate recommendations and better user experience. Our goal is different in that, the product of interest is explicit in the user query. Our goal is to *understand* the underlying user's intent, rather than to provide the most relevant recommendation. Nevertheless, advanced product embeddings used for product recommendation might help in capturing the customer affinity to a specific product and improve further the shopping need classification.

Behavioral Features in Search and Recommendation. The works in [1,21] propose the integration of implicit user feedback for ranking Web search results. Although these works tackle a different problem and are not comparable to our work, we share a similar goal in that we obtain user behavior features (e.g. purchase history) that can serve to learn a latent user shopping intent model.

Works on session based recommendation [19,29,32,37] propose approaches for incorporating diverse features to facilitate recommendation within a given user session (e.g. music search). At its core it predicts the next item with which the user will interact. These works are complementary to ours as they address the question on *what* to recommend, whereas we focus on determining *when* to trigger recommendations by the voice assistants recommender systems.

3 Data Analysis

3.1 Dataset and Labels

Since we cannot observe the true user intent, we rely on their purchase behavior: if the user asked a question and then purchased the referred product, we consider it as having a shopping need.

A shopping intent does not often result in an instantaneous purchase. Selecting the time window for the purchase information, from the time the user issued a PQ up to the actual purchase, represents an interesting tradeoff. Longer periods tend to better capture the hidden purchase intent, however at the same time pose the risk of labeling unrelated utterances with shopping intent. We analyzed different time windows, including one hour/day/week/two weeks, and 28 d, and found that a period of 28 d increased coverage by 187% while increasing noise by 8% only[3] compared to a period of one week. Given a PQ, we checked the future user purchases. If the referred product – or a similar product, sharing the same category – was purchased within 28 d, we label the PQ as SPQ, otherwise NSPQ.

We collected a dataset of 374k PQs from a leading voice assistant, evenly distributed among SPQs and NSPQs[4], split into 306k/34k/37k subsets for train/validation/test set, respectively. The different splits have non-overlapping

---
[3] We estimated noise by scrambling the utterances and purchases and measure how often an utterance is coincidentally followed by a purchase.
[4] The sample was designed to have an even distribution of NSPQs and SPQs, as such it is not representative of the overall traffic and none of the results reported here should be extrapolated to the entire traffic.

users, thus preventing intent data leakage. For each question, we collected its text, voice assistant's response, the product and its category, and user past purchases.

3.2 Data Insights

To understand correlated signals with SPQs, we perform a correlation analysis (Pearson's correlation coefficient), between the question type, their timing, and the searched product.

Questions. The relation between the user's shopping need and the question text is involved. We found that questions asked from users about vitamins or petfood have a high shopping intent (88%, 82%), while questions about televisions or cellphones have low shopping intent (only 16%, 18%). Figure 2 shows the histogram of the SPQ rate for the 50 most popular product categories. We see that different product types have different SPQ rates compared to the average 50%. There are product types where most of the queries are general knowledge need (4 categories with SPQ rate below 0.25%), while there are product types where most questions have a shopping need (5 categories with SPQ rate above 0.75%). For the rest, more than half of the popular categories have SPQ rate above 60%. This demonstrates that the product category, which can be inferred from the text of the question, is indicative of SPQs.

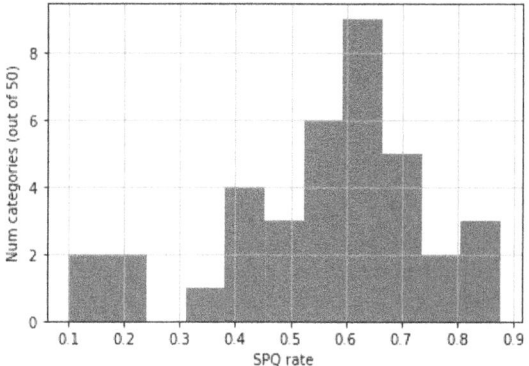

Fig. 2. SPQ rate per product type (out of 50 types with at least 1000 PQs, which represent 29.7% of the entire dataset).

However, while a shopping need is correlated with some textual features of the questions, this correlation is not natural nor intuitive. To demonstrate this, we sampled 10K questions and presented them to annotators. The annotator was presented with the question, the voice assistant response, and was required to estimate whether the user will likely purchase the referred product. To demonstrate this, we labelled 10K utterances, asking annotators to predict whether the user will purchase the product. We saw a very low correlation of 0.05 between human evaluation and future purchases. This indicates that from text alone, humans cannot predict whether a question has a shopping intent. An example of this phenomena, e.g., *"what do fish eat?"*, which to annotators seems like a general knowledge need question, while later purchases of fishing materials demonstrate the latent shopping need behind this question.

Users. The user properties also have an involved relation with the shopping intent. On the one hand, we found that shopping need is not a user property. We took two random PQs from each user, and measured the correlation between the first one being an SPQ and the second being a NSPQ. To avoid two questions referring to the same product category, which could lead to a false correlation, we de-duplicated PQs from the same user about the same product category. The correlation was only $r = 0.14$, showing that the same user may ask questions that have or do not have a shopping need.

On the other hand, we found a relation between the user's shopping history and the shopping need. For each PQ, we checked if the referred product was previously purchased by the user (28 d prior). This property has a strong correlation of $r = 0.67$ with the SPQ property, which is based on a future purchase. This shows that when the user asks about a product category that was previously purchased, they are more likely to purchase again.

While there is strong correlation between SPQs and preceding purchases, this might simply reflect user tendency to repurchase products, and be unrelated to the question. To refute this possibility, we measured the correlation of purchasing the referred product during two consecutive 28-days periods, before the actual user's question (PQ). We found that the correlation is significantly weaker when compared to consecutive periods coming before and after the user's PQ, showing that the PQ plays an important role. More specifically, given a PQ referring to product P from the user's shopping history, we consider four 28-days time periods T_0, T_{-1}, T_{-2} and T_{-3}, where T_0 is the time period coming after asking the question, T_{-1} is the time period coming before the question, and consecutively $T_{-3} < T_{-2} < T_{-1}$. The correlation between purchasing product P during T_0 and purchasing P during T_{-1} is measured to be $r = 0.67$. If this correlation was due to the periodic nature of purchasing P, we would expect a similar correlation of purchasing P also during T_{-2} and T_{-3}. However, the correlation was $r = 0.18$, showing that the high correlation of purchasing P between times T_0 and T_{-1} is strongly connected to the PQ, which indicates its shopping intent.

4 Identifying SPQs

To identify SPQs we present a model SPQI with three components: (a) features of different types, (b) feature aggregation through mixture of experts [45], and (c) a Graph Attention Network. Figure 3 presents a detailed overview of our

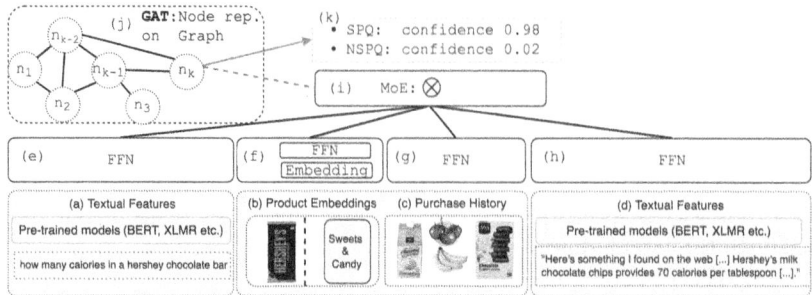

Fig. 3. SPQI: For an incoming product question SPQI performs the following steps to predict whether it is an SPQ: (a) computing the question text representation using pre-trained transformer models; (b) extracting the queried product (e.g. *"hershey chocolate bar"*); (c) extracting its category in the product catalog (e.g. *"sweets & candy"*); (d) encoding the voice assistant response using pretrained models as in (a). In (e) – (h) the input features are pushed through a FFN layer, allowing the features to be of the same dimensionality. In (i) the features are combined using the MoE approach, and are then fed into GAT's message-passing, which computes the node representation according to a computed sub-graph at batch time (j), each node representing a question. The resulting node representation in (k) is used for classifying the question as either SPQ or NSPQ using a Sigmoid layer.

approach in identifying questions with *shopping intent*. Our approach consists of two main components. First, a question encoder, which learns a joint question feature representation from different feature modalities. Second, the question representation is fed onto the GAT component, which for a constructed sub-graph, computes the final question representation used for identifying utterances with shopping need. In the following, we provide a detailed overview of the computed utterance features, and how they are utilized in our GAT based SPQ classification approach.

Textual Features. The PQ is an important intent indicator, containing key aspects that are asked about a product (e.g., *price, delivery time*, etc.). Additionally, voice assistant's response to the question contains information if the question was understood and answered. Furthermore, when a customer asks a question the voice assistant responds with either an *answer* or indicates that it *does not understand* the utterance. For cases where the voice assistant provides an answer, the user can continue their conversation by either asking additional questions regarding the product or make a shopping decision. We encode textual features extracted with a RoBERTa model [33], and use the [CLS] token for representation.

Product Features. An important aspect of a PQ is the queried product itself. Products have a rich structure, and users may have shopping preferences towards seemly unrelated products or specific categories, e.g., because they share a similar price range or a similar theme, e.g., *"zucchini"* and *"broccoli"*, or *"healthy food"*.

We use the product and the product category embeddings[5] that are trained from scratch to capture shopping need related features.

We compute the product p, product category c, and parent category $\hat{c}$ of c embeddings. The features are considered independently from each other[6] and allow the model to learn how to leverage them during training.

Behavioral Features. To identify shopping intent, the user's shopping behavioral patterns, extracted from the purchase history, are key. The purchase history is a sparse feature, consisting of the tuples ⟨user, product⟩. We pre-train a skip-gram based embeddings [4] for better generalization. We use the tuple ⟨u, p⟩ to train a model that predicts if the product p will be purchased by user u, where $u, p \in \mathbb{R}^{50}$ represent the product and user vectors. The resulting embedding have two highly desirable properties: (1) create user vectors that are similar according to purchasing patterns, and (2) product vectors that are similar according to the users that co-purchased them.

Using the pre-trained model, for an input question, from the list of purchases made by the user in previous 28 d **H**, we compute: (i) average dot/cosine similarity, maximum dot/cosine similarity, and the sum of dot/cosine similarity.

4.1 Learning Joint Question Representations via MoE

An important question is how to combine the various feature types (textual, numerical, categorical). Typically, such features are combined using either max or average pooling mechanisms [13, 52]. One drawback of this is that they do not allow the models to learn weights for each feature type according to their impact on the classification task.

We propose the use of mixture-of-experts [45], which allows the dynamic mixing of the different question feature types. Let $\mathbf{F} \in \mathbb{R}^{f \times n}$ be the concatenated set of features. For each feature dimension in $\mathbf{F}$, MoE weighs the importance of the f individual feature types as follows.

$$\text{MoE} = \mathbf{W}'_e \left(\mathbf{F}\mathbf{W}_e + c_1\right)^T + c_2; \qquad \lambda_{f_j, z} = \frac{\text{MoE}_{f_j, z}}{\sum_{f' \in |f|} \text{MoE}_{f', z}} \qquad (1)$$

where $\mathbf{W}_e \in \mathbb{R}^{n \times i}$, $\mathbf{W}'_e \in \mathbb{R}^{n \times i}$, represent trainable parameters used to compute feature type weights. Finally, $\lambda_{f_j, z} \in \mathbf{\Lambda}$ represents the feature weight for dimension z and feature type f_j, while $\mathbf{\Lambda} \in \mathbb{R}^{f \times n}$ represents all the feature weights.

The final joint question representation that is passed onto the GAT framework, is computed by performing the Hadamard product between each feature vector and the computed MoE weights: $\mathbf{h} = \sum_{f' \in |f|} \mathcal{F}_{f'} \circ \mathbf{\Lambda}_{\mathbf{f}'}$.

4.2 Question Graph Construction and Classification

Considering questions as independent instances has the disadvantage of not being able to leverage similar *questions* or *users* that may have similar purchasing

---
[5] Products are organized in a taxonomic product category graph.
[6] As future work we foresee integrating graph embeddings for categories and products.

behavior. Hence, we formulate the SPQ identification problem as a node classification task using GAT [50]. GAT allow to create soft dependencies between questions (if they share the same product), and as such, influences the prediction of shopping need for a question, by taking into account questions about the same product from other users. An important aspect of utilizing GATs, is the shape of the graph (i.e. node connectivity). As nodes are the user questions, represented by the joint representation **h**, and nodes are connected if they share the same queried *product*.

We train a GAT model to compute the question representations: $\widehat{\mathbf{h}}_q = \texttt{GAT}(\mathbf{h}_q)$, which is used to obtain the SPQ probability as $\mathbf{s}_q = \sigma(\widehat{\mathbf{h}}_\mathbf{q} \cdot \theta)$ to classify into either an SPQ or a NSPQ.

5 Experiments

We discuss the offline and online evaluation results, and further introduce baselines for the offline experimentation phase.

5.1 Baselines and Ablations

RoBERTa-Baseline. This baseline highlights the difficulty of identifying shopping intent from text alone. We consider two variations: (1) RoBERTa-Question, where only the question text is used for training and inference, and (2) RoBERTa-Text, where additionally voice assistant's answer is used for training and inference.

MLP-Baseline. All the computed features are pushed through an MLP layer for classification. We consider several ablations (MLP-Question and MLP-Text are identical to the RoBERTa baselines) of this baseline.

SPQI Setup and Ablations. We consider different model ablations using different feature subsets, and we distinguish between SPQI-MoE and SPQI-CONCAT, with the difference being on how the diverse features are aggregated with MoE or simply concatenated. SPQI has four layers of graph convolutions [23], and each node has 1024 dimensions.

6 Evaluation Results

6.1 Offline Evaluation

Model Performance. Table 1 shows the results of identifying SPQs. Our approach, SPQI-MoE achieves the best results with $F1 = 0.91$. SPQI-Concat-Full has a 1% drop in F1. This shows that the flexibility of MoE to dynamically decide per question, which features are important in predicting user's shopping need.

Comparing SPQI-Full-MoE against MLP-MoE-Full, the use of GATs provides an additional advantage in terms of F1. This shows that even for the same feature

Identifying Shopping Intent in Product QA for Proactive Recommendations

Table 1. SPQ detection performance of competing approaches. Significant results (as per paired t-test) are marked with * ($p\text{-value} < 0.05$). SPQI-MoE ablations are compared to MLP-MoE, and SPQI-Concat vs. MLP-Concat. For the sake of clarity, we omit from the table the results for RoBERTa-Question and RoBERTa-Text, given that they are the lowest performing with F1=0.694 and F1=0.699, respectively. On the same feature set, SPQI-MoE-Query and SPQI-MoE-Text obtain significantly higher results with F1=0.777 and F1=0.774, respectively.

Model	Behavior			Product			Text+Product			Text+Behavior			Product+Behavior			Full		
	P	R	F1	P	R	F1	P	R	F1	P	R	F1	P	R	F1	P	R	F1
MLP-Concat	0.827	0.910	0.866	0.758	0.755	0.757	0.758	0.766	0.762	0.856	0.916	0.885*	0.876	0.881	0.878	0.875	0.881	0.878
MLP-MoE	0.907	0.886	0.896	0.756	0.758	0.757	0.761	0.730	0.744	0.895	**0.921**	0.907*	0.903	0.908	0.906	0.903	0.914	0.908
SPQI-Concat	0.903	0.884	0.893*	0.760	0.757	0.759	0.760	0.772	0.766	0.898	0.893	0.895	0.890	0.904	0.895*	0.895	0.909	0.902*
SPQI-MoE	0.909	0.890	0.900*	0.762	0.764	0.763*	0.760	0.793	0.776*	**0.914**	0.885	0.900	0.903	0.910	0.907	0.903	0.917	**0.910***

set and way to combine them, GATs ability to consider neighbouring nodes for determining shopping need is helpful. This in itself is intuitive and has been widely studied in recommender systems (e.g. collaborative filtering), where tying users with similar search patterns can be helpful in recommendations. In our case, according to how subgraphs are constructed for classification, our models exploit information coming from users who share similar purchasing patterns and ask similar questions. Finally, RoBERTa baselines obtain the lowest performance, showing that identifying SPQs cannot be done from the question text alone.

In nearly all the cases, combining the question features using the MoE approach, allows the models to feature concatenation. Furthermore, computing the final question representation using GATs, where neighbouring nodes can have an impact on its representation, allows the model to obtain more accurate intent classification results.

Although the SPQI models are larger (in parameters) than their MLP counterparts, the performance decrease is not due to model size, as the comparison between SPQI-MoE and SPQI-Concat shows that SPQI-MoE achieves better performance.

Ablations. The user behavioral features achieve the biggest performance improvement. SPQI-Text+Behavior is the most accurate model with $P = 0.914$, which is 1% higher than SPQI-Full-MoE. However, in terms of recall we get a drop of 3%. The addition of product embedding features, allows the model to improve its recall by covering cases of users that may have a sparse purchase history or that few purchased products within the same category as the queried product. Interestingly, none of the feature combinations manage to achieve the full performance of SPQI-MoE, where we draw conclusions that while the precision may be higher on certain feature subsets (e.g. SPQI-MoE-Text+Behavior), the combination of all the features allows to obtain maximum coverage.

Similarly, here too, the SPQ-MoE and SPQ-Concat ablations perform better than most of their counterparts. In only one case the MLP obtain significantly better performance.

In terms of architecture, SPQI-MoE-Text obtains an increase of $F1 = +7\%$ over RoBERTa. This improvement can be attributed to the use of GATs. Additionally, we do not see any significant difference between the models when trained only on the users questions, as opposed to both the question and the voice assistant answers. One reason may be that answers are usually handled by Q&A systems that do not provide personalized answers to the users.

Overall, only MLP-MoE-Text+Behavior achieves results that are significantly better than those of SPQI-MoEText+Behavior. The performance decrease comes from lower recall.

In summary, the ablations show that our contributions are twofold. First, the proposed features that capture user's shopping need are highly suitable for voice search. Second, the proposed MoE for integrating the different features into a joint question representation, and the use of GATs for the intent classification task, with consistent improvement over the competing approaches (e.g., MLP-Concat/MoE). Note that the MLP baseline represents a strong baseline, since it is trained on the proposed features and uses the MoE.

6.2 Online Recommendation Evaluation

The SPQ definition in Sect. 3 is only a proxy of shopping intent. PQs can theoretically be unrelated to the associated purchase. Our online evaluation, with real voice assistant users, verifies that the identified SPQs by our SPQI-MoE have a shopping intent. For this experiment, we split the users into two groups: 1) $T1$ where the users, according to our model, are recommended a shopping action to add a product into their shopping list, and 2) C where all the users issuing a PQ are prompted to add the product into their shopping list.

$T1$ obtains 81.5% higher F1 score than C in identifying SPQs, and additionally users on $T1$ were 79.5% more likely to add the product into their shopping list when compared to users in C. This experiment shows that our approach is able to identify user's shopping need, validated by users adding the queried product into their shopping list. Note that, not adding a product to the shopping list does not indicate the contrary, namely, the lack of shopping intent.

7 Conclusion

We presented an approach to make proactive shopping recommendations in a leading industry voice assistant. By identifying shopping need, we allow the voice assistant to accurately recommend to its users relevant products or proactively suggest shopping actions that enhance their shopping experience.

We proposed a set of features that capture shopping intent from various perspectives, and relied on graph attention networks and on a novel mechanism to combine features for classifying a PQ as an SPQ. Experimental evaluations show that our proposed features and the way we encode user question through GATs, yield significant improvements over text classification approaches, achieving an increase in $F1$ of $+21\%$. The experiments confirm the highly latent nature of shopping intent. This work, unlike others, explores users' shopping intent in voice, not only in the context of product search, but also in the context of product question answering. Furthermore, focusing on informational intent and exploring whether the underlying need behind is related to general knowledge or to shopping.

References

1. Agichtein, E., Brill, E., Dumais, S.T.: Improving web search ranking by incorporating user behavior information. In: SIGIR (2006). https://doi.org/10.1145/1148170.1148177
2. Ahmadvand, A., Kallumadi, S., Javed, F., Agichtein, E.: Jointmap: joint query intent understanding for modeling intent hierarchies in e-commerce search. In: SIGIR (2020). https://doi.org/10.1145/3397271.3401184
3. Bi, K., Ai, Q., Croft, W.B.: Learning a fine-grained review-based transformer model for personalized product search. In: SIGIR (2021). https://doi.org/10.1145/3404835.3462911
4. Bojanowski, P., Grave, E., Joulin, A., Mikolov, T.: Enriching word vectors with subword information. Trans. Assoc. Comput. Linguistics **5**, 135–146 (2017). https://transacl.org/ojs/index.php/tacl/article/view/999
5. Carmel, D., Haramaty, E., Lazerson, A., Lewin-Eytan, L., Maarek, Y.: Why do people buy seemingly irrelevant items in voice product search?: On the relation between product relevance and customer satisfaction in ecommerce. In: WSDM (2020). https://doi.org/10.1145/3336191.3371780
6. Carmel, D., Lewin-Eytan, L., Maarek, Y.: Product question answering using customer generated content-research challenges. In: SIGIR (2018)
7. Chen, S., Li, C., Ji, F., Zhou, W., Chen, H.: Review-driven answer generation for product-related questions in e-commerce. In: WSDM (2019)
8. Chen, W., He, M., Ni, Y., Pan, W., Chen, L., Ming, Z.: Global and personalized graphs for heterogeneous sequential recommendation by learning behavior transitions and user intentions. In: RecSys (2022). https://doi.org/10.1145/3523227.3546761
9. Choi, J.I., et al.: Wizard of tasks: a novel conversational dataset for solving real-world tasks in conversational settings. In: Proceedings of the 29th International Conference on Computational Linguistics, pp. 3514–3529. International Committee on Computational Linguistics, Gyeongju, Republic of Korea (Oct 2022), https://aclanthology.org/2022.coling-1.310
10. Esmeli, R., Bader-El-Den, M., Abdullahi, H.: Towards early purchase intention prediction in online session based retailing systems. Electronic Markets, 1–19 (2020)
11. Faustini, P., Chen, Z., Fetahu, B., Rokhlenko, O., Malmasi, S.: Answering unanswered questions through semantic reformulations in spoken QA. In: Proceedings of the 61st Annual Meeting of the Association for Computational Linguistics (Volume 5: Industry Track), pp. 729–743. Association for Computational Linguistics, Toronto, Canada (Jul 2023). https://doi.org/10.18653/v1/2023.acl-industry.70, https://aclanthology.org/2023.acl-industry.70
12. Fetahu, B., Faustini, P., Castellucci, G., Fang, A., Rokhlenko, O., Malmasi, S.: Follow-on question suggestion via voice hints for voice assistants. In: EMNLP 2023 (2023)

13. Gao, J., Lyu, T., Xiong, F., Wang, J., Ke, W., Li, Z.: MGNN: a multimodal graph neural network for predicting the survival of cancer patients. In: SIGIR (2020). https://doi.org/10.1145/3397271.3401214
14. Gao, S., Ren, Z., Zhao, Y., Zhao, D., Yin, D., Yan, R.: Product-aware answer generation in e-commerce question-answering. In: Proceedings of WSDM, pp. 429–437 (2019)
15. Guo, L., Hua, L., Jia, R., Zhao, B., Wang, X., Cui, B.: Buying or browsing?: Predicting real-time purchasing intent using attention-based deep network with multiple behavior. In: SIGKDD (2019). https://doi.org/10.1145/3292500.3330670
16. Guy, I.: Searching by talking: Analysis of voice queries on mobile web search. In: Perego, R., Sebastiani, F., Aslam, J.A., Ruthven, I., Zobel, J. (eds.) SIGIR (2016). https://doi.org/10.1145/2911451.2911525
17. Hao, J., et al.: P-companion: a principled framework for diversified complementary product recommendation. In: CIKM (2020)
18. Haramati, E., Fairstein, Y., Lazerson, A., Lewin-Eytan, L.: External evaluation of ranking models under extreme position-bias. In: WSDM (2022)
19. Hidasi, B., Karatzoglou, A.: Recurrent neural networks with top-k gains for session-based recommendations. In: CIKM (2018). https://doi.org/10.1145/3269206.3271761
20. Ingber, A., Lazerson, A., Lewin-Eytan, L., Libov, A., Osherovich, E.: The challenges of moving from web to voice in product search. In: Proceedings of the 1st International Workshop on Generalization in Information Retrieval (GLARE 2018) (2018). http://glare2018.dei.unipd.it/paper/glare2018-paper5.pdf
21. Joachims, T.: Optimizing search engines using clickthrough data. In: SIGKDD
22. Khandokar, I.A., Islam, A.M., Islam, S., Shatabda, S., et al.: A gradient boosting classifier for purchase intention prediction of online shoppers. Heliyon **9**(4) (2023)
23. Kipf, T.N., Welling, M.: Semi-supervised classification with graph convolutional networks. In: ICLR (2017). https://openreview.net/forum?id=SJU4ayYgl
24. Kulkarni, A., Mehta, K., Garg, S., Bansal, V., Rasiwasia, N., Sengamedu, S.: Productqna: answering user questions on e-commerce product pages. In: Companion Proceedings of WWW (2019)
25. Lai, T., Bui, T., Li, S., Lipka, N.: A simple end-to-end question answering model for product information. In: Proc. of ECONLP, pp. 38–43 (2018)
26. Lai, T.M., Bui, T., Lipka, N., Li, S.: Supervised transfer learning for product information question answering. In: Proc. of ICMLA, pp. 1109–1114. IEEE (2018)
27. Le, D.T., Weber, V., Bradford, M.: Combining semantic search and twin product classification for recognition of purchasable items in voice shopping. In: 4th Workshop on e-Commerce and NLP (Aug 2021)
28. Le, T.H., Lauw, H.W.: Explainable recommendation with comparative constraints on product aspects. In: WSDM (2021)
29. Li, J., Ren, P., Chen, Z., Ren, Z., Lian, T., Ma, J.: Neural attentive session-based recommendation. In: CIKM (2017). https://doi.org/10.1145/3132847.3132926
30. Li, Y., Chen, W., Yan, H.: Learning graph-based embedding for time-aware product recommendation. In: CIKM (2017)
31. Ling, C., Zhang, T., Chen, Y.: Customer purchase intent prediction under online multi-channel promotion: a feature-combined deep learning framework. IEEE Access **7**, 112963–112976 (2019)
32. Liu, Q., Zeng, Y., Mokhosi, R., Zhang, H.: STAMP: short-term attention/memory priority model for session-based recommendation. In: SIGKDD (2018). https://doi.org/10.1145/3219819.3219950

33. Liu, Y., et al.: Roberta: A robustly optimized BERT pretraining approach. CoRR abs/ arXiv:1907.11692 (2019)
34. Lopatovska, I., et al.: Talk to me: exploring user interactions with the amazon alexa. J. Librariansh. Inf. Sci. **51**(4), 984–997 (2019)
35. Ma, R., Liu, N., Yuan, J., Yang, H., Zhang, J.: CAEN: a hierarchically attentive evolution network for item-attribute-change-aware recommendation in the growing e-commerce environment. In: RecSys (2022). https://doi.org/10.1145/3523227.3546773
36. McAuley, J., Yang, A.: Addressing complex and subjective product-related queries with customer reviews. In: Proceedings of of WWW, pp. 625–635 (2016)
37. Meng, W., Yang, D., Xiao, Y.: Incorporating user micro-behaviors and item knowledge into multi-task learning for session-based recommendation. In: SIGIR (2020). https://doi.org/10.1145/3397271.3401098
38. Rafailidis, D., Manolopoulos, Y.: The technological gap between virtual assistants and recommendation systems. CoRR abs/ arXiv:1901.00431 (2019)
39. Ramnath, K., Sari, L., Hasegawa-Johnson, M., Yoo, C.: Worldly wise (wow) - cross-lingual knowledge fusion for fact-based visual spoken-question answering. In: NAACL-HLT (2021). https://doi.org/10.18653/v1/2021.naacl-main.153
40. Ravichander, A., et al.: Noiseqa: Challenge set evaluation for user-centric question answering. In: EACL (2021). https://aclanthology.org/2021.eacl-main.259/
41. Rhee, C.E., Choi, J.: Effects of personalization and social role in voice shopping: An experimental study on product recommendation by a conversational voice agent. Comput. Hum. Behav. 109, 106359 (2020). https://doi.org/10.1016/j.chb.2020.106359
42. Rozen, O., Carmel, D., Mejer, A., Mirkis, V., Ziser, Y.: Answering product-questions by utilizing questions from other contextually similar products. In: Proceedings of NAACL, pp. 242–253 (2021)
43. Rzepka, C.: Examining the use of voice assistants: A value-focused thinking approach (2019)
44. Salle, A., Malmasi, S., Rokhlenko, O., Agichtein, E.: Studying the effectiveness of conversational search refinement through user simulation. In: Proceedings of 2021 European Conference on Information Retrieval (ECIR 2021), pp. 587–602. Springer (2021). https://doi.org/10.1007/978-3-030-72113-8_39
45. Shazeer, N., et al.: Outrageously large neural networks: The sparsely-gated mixture-of-experts layer. In: ICLR (2017), https://openreview.net/forum?id=B1ckMDqlg
46. Shen, Y., Yan, J., Yan, S., Ji, L., Liu, N., Chen, Z.: Sparse hidden-dynamics conditional random fields for user intent understanding. In: WWW (2011). https://doi.org/10.1145/1963405.1963411
47. Tagliabue, J., Yu, B., Bianchi, F.: The embeddings that came in from the cold: Improving vectors for new and rare products with content-based inference. In: RecSys (2020). https://doi.org/10.1145/3383313.3411477
48. Tyagi, A., et al.: Fast intent classification for spoken language understanding systems. In: IEEE ICASSP (2020). https://doi.org/10.1109/ICASSP40776.2020.9054240
49. Vasile, F., Smirnova, E., Conneau, A.: Meta-prod2vec: product embeddings using side-information for recommendation. In: RecSys (2016). https://doi.org/10.1145/2959100.2959160
50. Velickovic, P., Cucurull, G., Casanova, A., Romero, A., Liò, P., Bengio, Y.: Graph attention networks. In: ICLR (2018). https://openreview.net/forum?id=rJXMpikCZ

51. Wang, J., Sarwar, B., Sundaresan, N.: Utilizing related products for post-purchase recommendation in e-commerce. In: RecSys (2011). https://doi.org/10.1145/2043932.2043995
52. Wei, Y., Wang, X., Nie, L., He, X., Hong, R., Chua, T.: MMGCN: multi-modal graph convolution network for personalized recommendation of micro-video. In: ACM MM. ACM (2019)
53. Xu, H., Xie, S., Shu, L., Philip, S.Y.: Dual attention network for product compatibility and function satisfiability analysis. In: AAAI, vol. 32 (2018)
54. Yang, J., Drake, T., Damianou, A.C., Maarek, Y.: Leveraging crowdsourcing data for deep active learning an application: Learning intents in alexa. In: WWW (2018). https://doi.org/10.1145/3178876.3186033
55. Yu, J., Zha, Z.J., Chua, T.S.: Answering opinion questions on products by exploiting hierarchical organization of consumer reviews. In: Proceedings of EMNLP, pp. 391–401 (2012)
56. Zhang, S., Lau, J.H., Zhang, X., Chan, J., Paris, C.: Discovering relevant reviews for answering product-related queries. In: ICDM 2019 (2019)
57. Zhao, J., Chen, H., Yin, D.: A dynamic product-aware learning model for e-commerce query intent understanding. In: CIKM (2019)
58. Zheng, L., Noroozi, V., Yu, P.S.: Joint deep modeling of users and items using reviews for recommendation. In: WSDM (2017). https://doi.org/10.1145/3018661.3018665

KGUF: Simple Knowledge-Aware Graph-Based Recommender with User-Based Semantic Features Filtering

Salvatore Bufi[1]([✉]), Alberto Carlo Maria Mancino[1], Antonio Ferrara[1], Daniele Malitesta[2], Tommaso Di Noia[1], and Eugenio Di Sciascio[1]

[1] Politecnico di Bari, Bari, Italy
{salvatore.bufi,alberto.mancino,antonio.ferrara,tommaso.noia, eugenio.sciascio}@poliba.it
[2] Université Paris-Saclay, CentraleSupélec, Inria, Orsay, France
daniele.malitesta@centralesupelec.fr

Abstract. The recent integration of Graph Neural Networks (GNNs) into recommendation has led to a novel family of Collaborative Filtering (CF) approaches, namely Graph Collaborative Filtering (GCF). Following the same GNNs wave, recommender systems exploiting Knowledge Graphs (KGs) have also been successfully empowered by the GCF rationale to combine the representational power of GNNs with the semantics conveyed by KGs, giving rise to Knowledge-aware Graph Collaborative Filtering (KGCF), which use KGs to mine hidden user intents. Nevertheless, empirical evidence suggests that computing and combining user-level intent might not always be necessary, as simpler approaches can yield comparable or superior results while keeping explicit semantic features. Under this perspective, user historical preferences become essential to refine the KG and retain the most discriminating features, thus leading to concise item representation. Driven by the assumptions above, we propose KGUF, a KGCF model that learns latent representations of semantic features in the KG to better define the item profile. By leveraging user profiles through decision trees, KGUF effectively retains only those features relevant to users. Results on three datasets justify KGUF's rationale, as our approach is able to reach performance comparable or superior to SOTA methods while maintaining a simpler formalization.

Keywords: recommendation · knowledge graphs · graph neural networks

1 Introduction

Recommender Systems find widespread application across diverse domains, owing to their proficiency in furnishing users with personalized items [10, 46, 48].

D. Malitesta—Work done while at Politecnico di Bari.

Typically, these systems rely on past user-item interactions to extrapolate new preferences and generate personalized suggestions [12,20]. Collaborative Filtering [24] (CF) methods adeptly extract similarity patterns from user-item interactions, constituting (to date) the most diffused and winning recommendation paradigm [20,27,29]. Nevertheless, by design, the learned user and item profiles in traditional CF recommender systems cannot embed the information conveyed by the underlying multi-hop user-item relations, thus disregarding a rich source of potentially-useful information [28,40]. Following the popular graph representation learning [18] wave, Graph Neural Networks (GNNs) have recently been applied to CF approaches, giving rise to a fresh family of recommender systems [19,28,44,47], better known as Graph Collaborative Filtering (GCF). Despite the success of traditional CF and GCF recommendation systems, such approaches often overlook the nuances in user preferences related to specific item characteristics; indeed, this represents a missed opportunity for enhancing user personalization. As opposed to CF systems, content-based recommendation [15] leverages the attributes of historically-favored products, proposing unseen items that align with user distinct tastes [20]. In these systems, Knowledge Graphs (KGs) have been shown to provide a meaningful source of structured information [4,11,49,50] used to craft features that enrich the item description, thanks to the mapping between items and semantic entities, and helps to tackle open challenges in recommendation, such as the graph sparsity.

As a recent trend, an increasing number of recommender system algorithms currently propose to combine the representational capabilities of GNNs with the rich information conveyed by KGs, resulting in a novel hybrid family of recommendation techniques which might be suitably named as Knowledge-aware Graph Collaborative Filtering [3,21,41,45,49] (KGCF). Most of KGCF methods model the user intent to enhance recommendation quality and interpretability, thus outperforming similar baselines techniques [26,39,41]. However, such architectures often require convoluted aggregations of several ad-hoc methodologies and strategies to learn meaningful user and item profiles, making it hard to apply KGCF approaches in real-world scenarios. In this regard, we believe that these architectures can be dramatically simplified while maintaining comparable or even superior performance. In particular, we consider to utilize the content information from KGs solely for describing item representations, refraining from any attempt to directly incorporate it into building user profiles, which can be instead indirectly learned by means of collaborative filtering techniques.

Driven by these assumptions, this paper presents a novel KGCF recommendation algorithm named **KGUF** (short for **K**nowledge **G**raph **U**ser-based **F**iltering) designed to efficiently harness side information from KGs. From a technical perspective, KGUF learns user and item embeddings through linear propagation on the collaborative graph. In contrast to other approaches, our solution, sketched in Fig. 3, is able to integrate latent representations for semantic features right into the item representation. Moreover, sharing the same principle but replacing it with a simple mechanism, we propose substituting the intent modeling with a decision tree mechanism that selects the most meaningful semantic features in the KGs and filters out noisy features that poorly

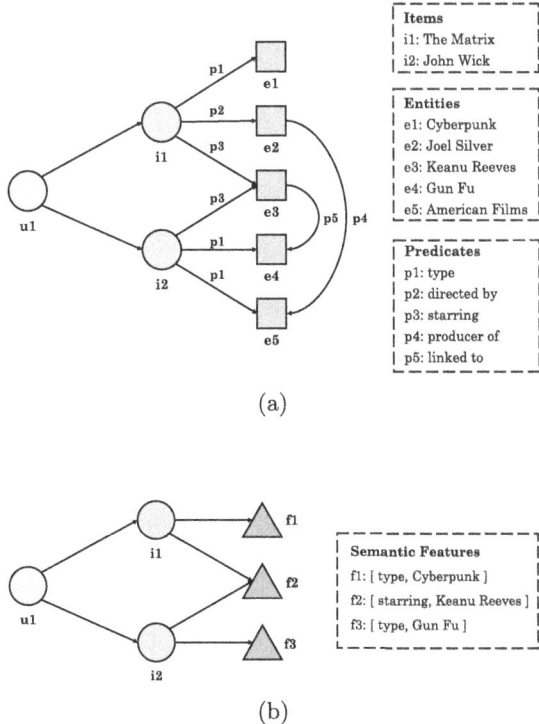

Fig. 1. An example of how KGUF models the item representation. Figure 1a shows the entities in the knowledge graph linked to the items. In Fig. 1b, the most relevant semantic features selected by KGUF are reported.

contribute to the item description but heavily impact the computational costs. In our work, we take inspiration from the KGTORE method [26] and employ a similar technique to identify the most influential user features, thereby eliminating the need to process the entire knowledge graph. However, our approach differs from KGTORE in that we do not explicitly incorporate the reasoning behind user-item interactions into our model. This results in a more lightweight and efficient methodology.

All in all, the key contributions of this paper are summarized as follows:

1. We propose a novel recommendation model, namely KGUF, which effectively selects and integrates user-relevant semantic features during the graph learning phase to improve item representation.
2. We conduct extensive experiments over three well-known recommendation datasets to assess the efficacy of our framework and evaluate the contribution of collaborative and content signals on KGUF.
3. We delve into using decision trees for selecting semantic features relevant to users, investigating how the effects of negative sampling and the presence of constraints regarding the tree depth affect the recommendation performance.

To foster the replicability of the experiments and results, we release the code for KGUF and the baselines, along with the datasets and configurations[1].

2 Related Work

This section aims to survey and compare other approaches in recommendation literature leveraging KGs and graph representation learning strategies.

2.1 KGs Information in Recommendation

As a general trend, the incorporation of information coming from KGs into recommendation systems has consistently demonstrated improvements in model performance and interpretability across diverse domains [4,11,49,50]. Indeed, the utilization of KGs has been effective in boosting different recommendation tasks, including hybrid collaborative/content-based recommendation [4,25], where the KG is of support to the lack of collaborative information, eventually enhancing performance; explainable recommendation [6,35], where the KG serves as a robust information source for delivering human-like explanations; and user modeling [5,38], with the semantic description of items contributing to defining user profiles. KGs also offer a solution to the cold-start problem by augmenting observed interactions with KG-reachable unobserved items [36]. However, it is crucial to note that in some cases, improper extraction of information may potentially weaken the model capabilities [9].

2.2 Graph Representation Learning for Recommendation

Recommender systems based on graph representation learning strategies have demonstrated their superior performance by mining high-order user-item relationships embedded within the graph structure [30,31,34,44]. Inspired by the graph convolutional network layer (GCN) [22], early approaches like PinSage [46] and NGCF [40] have adapted GCN to the user-item bipartite graph. Later research has explored simplifying the GCN layer, as seen in methods like LightGCN [19] and LR-GCCF [8], which streamline model complexity by eliminating non-linearities and feature transformations in node embeddings. Additionally, UltraGCN [28] proposes a mathematical proxy to the infinite message-passing layer, thus limiting the negative effects of over-smoothing. Another avenue of research focuses on pruning noisy user-item interactions to discern meaningful preference patterns. Inspired by the graph attention network [37] (GAT), methodologies like DGCF [42] aim to identify the underlying intents of individual user-item interactions on a fine-grained level. Furthermore, works such as [2] prove how user-item edges can incorporate multimodal content, such as user reviews, to enhance recommendation and address the challenge of over-smoothing.

---

[1] https://github.com/sisinflab/KGUF.

2.3 KGs Information and Graph Representation Learning for Recommendation

Recent literature [21,45,49] has ventured beyond traditional collaborative filtering graph-based methods by incorporating information from KGs to enhance item descriptions. One example is KGAT [39], which explicitly models the KG in an end-to-end manner. It refines node embeddings using an attention mechanism to discern the importance of neighbors. In a different approach, the authors of KGIN [41] utilize the KG to define users' intents as weights, combining these weights with their relations. CKAN [43] proposes a method that integrates collaborative signals with knowledge data. More recently, KGs have been employed to formulate first-order logical rules, combined with graph embeddings, enabling recommenders based on neuro-symbolic reasoning [32]. Adding to this evolving landscape, in KGTORe [26] the authors propose an explicit interpretation of the edge between the user and the item, which is accounted for to drive the training of a GNN-based recommender system. Unlike previous models, KGTORe does not need to process the entire KG at training time but selects the most discriminative features for users, enhancing performance and personalization. Drawing inspiration from KGTORe, we employ a similar mechanism to pinpoint the most discriminative features for users, thereby sidestepping the necessity of processing the entire knowledge graph. Nonetheless, our methodology diverges from KGTORe by avoiding explicit integration of the rationale behind user-item interaction in our model, leading to a lighter and more efficient approach.

3 Background

In this section, we introduce the notion of KGs and the basics of decision trees used for content modelling, along with the problem we aim to address.

3.1 Knowledge Graphs for Recommendation

A knowledge graph stores structured information of real-world facts in the form of a heterogeneous graph which can be used to retrieve item attributes [7,14]. Formally, it is presented as a set of triples in the form $\sigma \xrightarrow{\rho} \omega$. Each triple indicates that the head entity σ is connected to the tail entity ω through the relation ρ. Given a set of entities $\mathcal{V}$ and a set of relations $\mathcal{R}$, a knowledge graph $\mathcal{KG}$ can be represented as follows:

$$\mathcal{KG} = \{\sigma \xrightarrow{\rho} \omega \mid \sigma, \omega \in \mathcal{V}, \rho \in \mathcal{R}\}. \tag{1}$$

Mapping items in the catalog using knowledge graphs, i.e., $\mathcal{I} \subseteq \mathcal{V}$, enables the descriptions of items based on semantic features.

3.2 Decision Trees

A decision tree is a hierarchical tree-like structure used in machine learning for classification tasks. It iteratively partitions the dataset into subsets depending on the values of input features, resulting in a tree structure in which each internal node represents a feature-based decision. The decision-making process is guided by criteria such as Gini impurity or information gain.

In practical terms, decision trees aim to identify variables whose values contribute the most to reducing another random variable's entropy, or uncertainty. Mathematically, the entropy $H(V)$ of a random variable V with k possible values in $v_1, ..., v_k$ is formally defined as:

$$H(V) = -\sum_{i=1}^{k} \Pr(V = v_i) \log_2 \Pr(V = v_i). \tag{2}$$

If observing the value of another variable X with n possible values in $x_1, ..., x_n$ results in a reduction of the entropy of V, then X is said to exhibit information gain over V:

$$IG(V, X) = H(V) - H(V|X),$$
$$H(V|X) = \sum_{i=1}^{n} P(X = x_i) H(V|X = x_i). \tag{3}$$

Decision trees are one possible lightweight solution to filter out noisy semantic features in the KG and select the most relevant ones from the user perspective.

4 Methodology

In this section, we present KGUF and its components. In particular, we delve into the semantic item modeling and its refinements based on user decisions. Then, we introduce our simple but effective knowledge-aware aggregation schema.

4.1 Modeling Items Through Knowledge Graph

Assuming that each item in the item set $\mathcal{I}$ of our recommendation problem is associated with an entity in $\mathcal{V}$, essentially being one of the entities within the knowledge graph ($\mathcal{KG}$), we can model the items in the catalog by leveraging the knowledge embedded in the KG. The rationale behind this modeling lies in the notion that each item can be comprehensively characterized by a distinct set of features, some of them influencing user decisions in varying ways.

Given this premise, the KG can be explored to identify connections with other entities for each item, serving as its attributes in the form $\sigma \xrightarrow{\rho} \omega$, where ρ and ω represent the predicate and object of the KG triple, respectively. As an example, if properly encoded in the reference knowledge graph, the movie *Harry Potter* could be semantically described by the predicate-object pairs ⟨*director, Chris Columbus*⟩ and ⟨*genre, Fantasy*⟩. Formally, we utilize the set of semantic features to depict the item $i \in \mathcal{I}$ based on its first-order connectivity extracted from the $\mathcal{KG}$:

$$\mathcal{F}_i = \{\langle \rho, \omega \rangle \mid i \xrightarrow{\rho} \omega \in \mathcal{KG}\}. \tag{4}$$

4.2 Refining Item Features with User Decision

When confronted with a novel item, users focus on a subset of its numerous features. To capture this user behavior, KGUF uses decision trees to conduct a behavioral analysis at the user level, leveraging her past ratings. The objective is to discern the most pertinent features within the knowledge graph that effectively describe items in the catalog while filtering out noisy ones. This tailored approach ensures that the model accentuates features that hold significant influence across all users in the system.

Formally, let $\mathcal{I}_u^+ \subseteq \mathcal{I}$ denote the set of items positively rated by each user u in the set $\mathcal{U}$ of our recommendation problem, and let $\mathcal{I}_u^- = \mathcal{I} \setminus \mathcal{I}_u^+$ be the set of items not enjoyed by user u. Let $\tilde{\mathcal{I}}_u^- \subseteq \mathcal{I}_u^-$ be a selection of items randomly sampled from $\mathcal{I}_u^-$, such that $|\tilde{\mathcal{I}}_u^-| = |\mathcal{I}_u^+|$. Finally, $\mathcal{I}_u = \mathcal{I}_u^+ \cup \tilde{\mathcal{I}}_u^-$ is the set of items selected to depict the past behavior of the user.

For each user u, the set of semantic features $\mathcal{F}_u = \bigcup_{i \in \mathcal{I}_u} \mathcal{F}_i$ is used to compute the user decision tree T_u that, accordingly to the definition of Information Gain [4], is able to find the minimum subset of semantic features $\mathcal{F}_u^{T_u} \subseteq \mathcal{F}_u$ that can distinguish whether an item belongs to $\mathcal{I}_u^+$ or $\tilde{\mathcal{I}}_u^-$. Finally, each item i is formally described by the subset of its semantic features that appear in at least one user decision tree:

$$\mathcal{F}_i^* = \mathcal{F}_i \cap \bigcup_{u \in \mathcal{U}} \mathcal{F}_u^{T_u}. \tag{5}$$

4.3 Knowledge-Aware Aggregation

Inspired by LightGCN [19], the embeddings of each user and item, denoted as $\mathbf{e}_u \in \mathbb{R}^d$ and $\mathbf{e}_i \in \mathbb{R}^d$ respectively, are acquired through the recursive aggregation of their multi-hop neighbors' representations via message propagation across L convolutional layers. The resulting representations at various levels, $(\mathbf{e}_u^{(0)}, \ldots, \mathbf{e}_u^{(L)})$ and $(\mathbf{e}_i^{(0)}, \ldots, \mathbf{e}_i^{(L)})$, emerge from the integration of content information with collaborative signals originating from first-hop neighbor nodes and propagated from the preceding layer.

By expressing knowledge graph features through latent vectors $\mathbf{e}_f$, KGUF adeptly represents items based on their explicit characteristics. This modeling approach furnishes an explicit content-based representation for each item, situating it within the latent feature space as an amalgamation of vectors delineating its distinct characteristics. Formally:

$$\begin{aligned} \mathbf{e}_u^{(l)} &= \sum_{i \in \mathcal{N}(u)} \frac{1}{\sqrt{|\mathcal{N}(u)|}\sqrt{|\mathcal{N}(i)|}} \mathbf{e}_i^{(l-1)}, \\ \mathbf{e}_i^{(l)} &= \frac{\alpha}{|\mathcal{F}_i^*|} \sum_{f \in \mathcal{F}_i^*} \mathbf{e}_f + (1-\alpha) \sum_{u \in \mathcal{N}(i)} \frac{1}{\sqrt{|\mathcal{N}(u)|}\sqrt{|\mathcal{N}(i)|}} \mathbf{e}_u^{(l-1)}, \end{aligned} \tag{6}$$

where α serves as a weight factor for balancing content and collaborative contributions in item representations.

Noticeably, in contrast to LightGCN, KGUF introduces a knowledge-aware correction term in the item embeddings, which will affect also user representation through message passing schema. The collaborative and content signals collaboratively contribute to capturing long-range influences and leveraging the rich information derived from knowledge graphs. The use of decision trees to mediate content information guarantees that only relevant features engage in embedding modeling, ensuring a succinct and meaningful representation of users and items.

After completing L propagation hops, the user and item embeddings are derived as follows:

$$\mathbf{e}_u^* = \sum_{l=0}^{L} \frac{1}{1+l} \mathbf{e}_u^{(l)},$$
$$\mathbf{e}_i^* = \sum_{l=0}^{L} \frac{1}{1+l} \mathbf{e}_i^{(l)}. \qquad (7)$$

The introduction of the scaling factor $1/(1+l)$ serves to mitigate the oversmoothing problem that may arise with an increased number of explored hops [19].

4.4 Model Learning

The training stage involves learning the values of user, item, and feature embeddings to align user-item estimated interactions with the actual preferences of users. Given the embeddings $\mathbf{e}_u^*$ and $\mathbf{e}_i^*$, we employ the dot product of user and item representations as the prediction, expressed as:

$$\hat{r}_{ui} = \mathbf{e}_u^{*\top} \mathbf{e}_i^*. \qquad (8)$$

To optimize the model parameters, we utilize the pairwise optimization algorithm Bayesian Personalized Ranking (BPR). This algorithm operates on the assumption that a user u favors a consumed item i^+ over a non-consumed item i^-. It optimizes the model by maximizing, for each pair of i^+ and i^-, a function based on the difference $r_{ui^+} - r_{ui^-}$. Specifically, by constructing a training set $\mathcal{T} = \{(u, i^+, i^-) \mid (u, i^+) \in \mathcal{R}, (u, i^-) \notin \mathcal{R}, i^- \in \mathcal{I}\}$, BPR aims to optimize the following loss, where σ represents the sigmoid function:

$$\mathcal{L}_{\text{BPR}} = \sum_{(u,i^+,i^-) \in \mathcal{T}} -\ln \sigma(\hat{r}_{ui^+} - \hat{r}_{ui^-}), \qquad (9)$$

Finally, we use a joint learning approach to include the L2-regularization loss as follows:

$$\mathcal{L} = \mathcal{L}_{\text{BPR}} + \lambda \|\Theta\|_2^2, \qquad (10)$$

here Θ encompass all the learnable parameter, specifically $\Theta = \{\mathbf{e}_u, \mathbf{e}_i, \mathbf{e}_f \mid u \in \mathcal{U}, i \in \mathcal{I}, f \in \bigcup_{i \in \mathcal{I}} \mathcal{F}_i^*\}$, while λ controls the regularization term.

Table 1. Statistics of tested datasets.

Dataset	# Users	# Items	# Interactions	Sparsity
Facebook Books	1,398	2,933	18,978	99.54%
Yahoo! Movies	4,000	2,626	69,846	99.34%
Movielens 1M	6,040	3,706	1,000,209	95.53%

5 Experimental Settings

This section delineates the experimental configuration employed to evaluate the recommendation performance of KGUF.

5.1 Datasets and Knowledge Graph

We compare the performance of KGUF with state-of-the-art solutions on three well-known recommendation datasets: *MovieLens 1M*[2], *Yahoo! Movies*[3], and *Facebook Books*[4]. MovieLens 1M and Yahoo! Movies are two datasets in the movie recommendation domain that collect explicit ratings (ranging from 1 to 5) from 6,040 and 4,000 users, respectively, over catalogs of 3,706 and 2,626 items. Overall, they report 1,000,209 and 69,846 ratings, respectively. *Facebook Books* is a dataset that collects 18,978 implicit user feedback in the book recommendation domain. Since its ratings refer to 1,398 users and 2,933 items, it is characterized by greater sparsity. All the datasets are evaluated in an implicit feedback scenario: explicit ratings are binarized by applying a threshold value of 3. Only ratings greater or equal to the threshold value are recorded as implicit positive feedback. In order to reduce the noisy contribution of users and items with too few occurrences, we apply an iterative 5-core strategy, dropping users and items with less than five appearances. Through explicit linkings, the items in the dataset catalogs are directly linked to their semantic representation in *DBpedia*, a well-recognized public knowledge graph. For a fair comparison, we filter the noisy item-linked entities following, for each model, the filtering approach reported by the original authors. The overall statistics of the tested datasets are reported in Table 1.

5.2 Baselines

We empirically validate the effectiveness of KGUF by comparing its performance in terms of recommendation accuracy with different baselines from five related categories described in the following.

---

[2] MovieLens 1M Dataset: https://grouplens.org/datasets/movielens/1m/.
[3] Yahoo! Movies User Ratings and Descriptive Content Information, v.1.0: https://webscope.sandbox.yahoo.com/catalog.php?datatype=r.
[4] Facebook Books (Linked Open Data challenge co-located with ESWC 2015): https://2015.eswc-conferences.org/important-dates/call-RecSys.html.

- **Unpersonalized.** We test Random and Most Popular recommendations as a basis for discerning significance and behaviors influenced by popularity.
- **Collaborative Filtering.** BPR-MF [29] and Item-kNN [23] are two widely recognized collaborative approaches. The former follows a pair-wise learning approach as KGUF, and the latter exploits similarities between items by computing the nearest neighbors.
- **Knowledge-Aware.** Among these models, we selected KaHFM [5] and KGFlex [4,13], since they share with KGUF the feature representation but differ in the way they exploit it. KaHFM uses knowledge for initializing latent factors in factorization machines. KGFlex, instead, models the user-item interaction by means of a sparse collection of the most relevant semantic features.
- **Graph-Based.** LightGCN [19] rethinks the classic GCN aggregation schema by discarding feature transformation and non-linear activation functions. It results in linear aggregations with improved performance. DGFC [42] introduces a graph disentangling strategy for intercepting hidden user intents to model intent-dependent user-item interactions.
- **Graph-based with Knowledge.** KGIN [41] exploits the KG for computing users' intents as a weighted combination of semantic relations. User-item connections are interpreted by means of a limited set of relevant intents. KGTORe [26] mines the user motivations by looking at the discriminant semantic features in her history. These explicit features are then used to enrich the user and the item node representation.

5.3 Evaluation Protocol and Reproducibility

For the sake of reproducibility, we provide our code, data, and comprehensive instructions to enable the replication of all stages in our experimental procedures[5]. This paper follows the same evaluation protocol and experimental setting of [26]. Moreover, in order to compare KGUF with the same results, we adopt the reproducibility framework Elliot [1]. Each dataset is split into train, test, and validation sets by retaining 72%, 20%, and 8% of the original user ratings, respectively [17], and adopting the *all unrated protocol* [33].

Coherently with [26], the embedding size is fixed to 64, and each vector is randomly initialized using a Xavier initializer [16]. Then, we train KGUF on 20 different combinations of hyperparameters explored with a Bayesian optimization algorithm and apply early-stopping with a 5-epoch patience. We validate the our model on nDCG@10, since this metric enhances the capability of predicting top-ranking items. Moreover, the batch size is adapted proportionally to the dataset size: 64, 256, and 2048, respectively, for *Facebook Books*, *Yahoo! Movies*, and *Movielens 1M*.

---

[5] https://github.com/sisinflab/KGUF.

6 Results and Discussion

This section examines the performance of KGUF when compared to state-of-the-art baselines. Additionally, we present hyperparameter and ablation studies conducted to evaluate the effectiveness of the proposed approach. Specifically, this section delves into experimental analyses aimed at answering the following research questions:

- **RQ1** *Overall performance:* How does KGUF performance compare to that of SOTA recommenders? Additionally, what is the impact of the proposed knowledge-aware aggregation schema on the overall performance?
- **RQ2** *Decision Tree Construction:* To what extent does the construction of trees impact the performance of the proposed framework?

6.1 RQ1: Overall Performance

Firstly, we compare the recommendation accuracy of all the methods on three different datasets and report the results in Table 2. We denote the highest-performing result by boldfacing it and underscore the second-best result.

We observe that KGUF shows performance comparable or superior to all baselines over the three datasets. This improvement can be attributed to two key factors. Firstly, by employing user-based filtering, KGUF selectively learns the most pertinent characteristics for describing items in the catalog. In other words, it effectively filters knowledge information that proves genuinely valuable for the recommendation task.

Secondly, the linear aggregation schema adeptly combines collaborative and knowledge signals, thus improving both item and user representations. Another observation is that GNN-based techniques (LightGCN, DGCF, KGIN, KGTORe) consistently outperform non-graph-based ones, underscoring the efficacy of these architectures in leveraging long-term user-item relationships. The superior performance of knowledge-aware GNN-based methods is particularly noteworthy, highlighting how graph models effectively incorporate and exploit knowledge graphs.

Finally, we observe that not all knowledge-aware models (KGFlex, kaHFM) outperform pure collaborative approaches, emphasizing how merely utilizing knowledge-aware information may not necessarily lead to improved performance.

To delve deeper into assessing the efficacy of the proposed aggregation schema, we analyze how varying the weight assigned to the knowledge-aware component within the KGUF impacts the outcome on the dataset *Yahoo! Movies*. The results in Fig. 2a involve varying the weight of α in Eq. 6 across a spectrum from 0.2 to 0.8, i.e., increasing the contribution of knowledge on the collaborative information. The results show that searching for the best balance between content and collaborative contributions is necessary, which in this case is 40% content and 60% collaborative. However, it is even more interesting to question the collaborative and the knowledge signals' separate impact when switched on/off. Performance in Fig. 2b makes it clear that removing the knowledge information affects the model

Table 2. Accuracy performance comparison of KGUF with the selected baselines. All the models are grouped into their respective categories.

Dataset	Model	nDCG	HR	Recall
Facebook Books	Random	0.0074	0.0256	0.0110
	MostPop	0.0787	0.2535	0.1117
	BPR-MF	0.0945	0.2973	0.1358
	Item-kNN	0.1039	0.3353	0.1556
	KGFlex	0.0712	0.2235	0.1001
	kaHFM	0.0843	0.2776	0.1221
	LightGCN	0.0969	0.3170	0.1435
	DGCF	0.1001	0.3192	0.1444
	KGIN	0.1023	0.3112	0.1448
	KGTORe	**0.1188**	**0.3601**	**0.1698**
	KGUF	<u>0.1156</u>	<u>0.3594</u>	<u>0.1697</u>
Yahoo! Movies	Random	0.0065	0.0342	0.0090
	MostPop	0.1185	0.3238	0.1501
	BPR-MF	0.2303	0.5228	0.2843
	Item-kNN	0.2258	0.5033	0.2658
	KGFlex	0.1758	0.4360	0.2022
	kaHFM	0.2085	0.5064	0.2621
	LightGCN	0.2230	0.5085	0.2712
	DGCF	0.2356	0.5308	0.2848
	KGIN	0.2368	0.5500	<u>0.2971</u>
	KGTORe	<u>0.2471</u>	<u>0.5566</u>	0.2944
	KGUF	**0.2561**	**0.5685**	**0.3058**
MovieLens 1M	Random	0.0096	0.0836	0.0036
	MostPop	0.1730	0.6537	0.0773
	BPR-MF	0.3100	0.8811	0.1627
	Item-kNN	0.2932	0.8504	0.1490
	KGFlex	0.0111	0.1058	0.0053
	kaHFM	0.2992	0.8654	0.1526
	LightGCN	0.3081	0.8813	0.1614
	DGCF	0.3124	0.8819	0.1637
	KGIN	0.3082	0.8854	0.1638
	KGTORe	<u>0.3212</u>	<u>0.8957</u>	<u>0.1710</u>
	KGUF	**0.3277**	**0.8998**	**0.1755**

capabilities harder than removing the collaborative one. This shows how KGUF benefits from semantic information and how combining the two signals gives superior performance to considering only one.

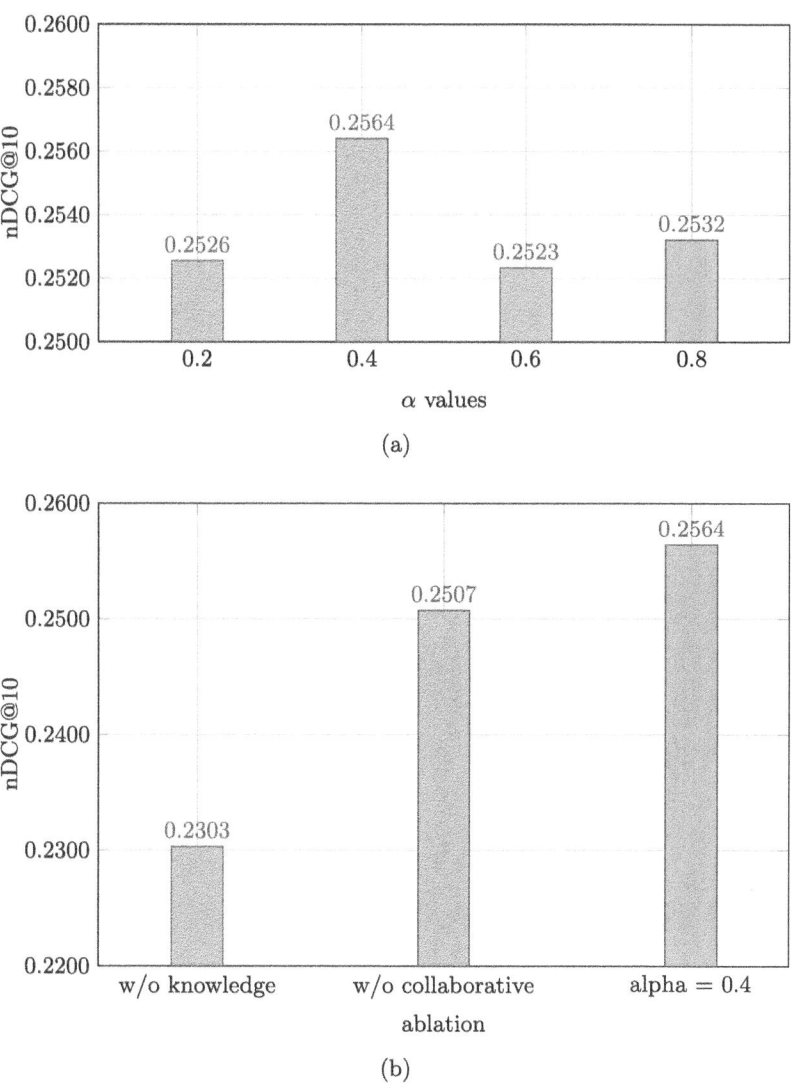

Fig. 2. Performance of KGUF regarding nDCG@10 when varying the knowledge contribution. Figure 2a reports the impact of the parameter α, which weights the knowledge contribution. Figure 2b shows the impact of considering the knowledge or the collaborative signals only.

6.2 RQ2: Decision Tree Construction

In this study, we explore how the tree construction strategies can impact the overall performance of KGUF, assessing how the negative sampling augmentation for tree building and the maximum depth of the trees contribute to the overall recommendation quality.

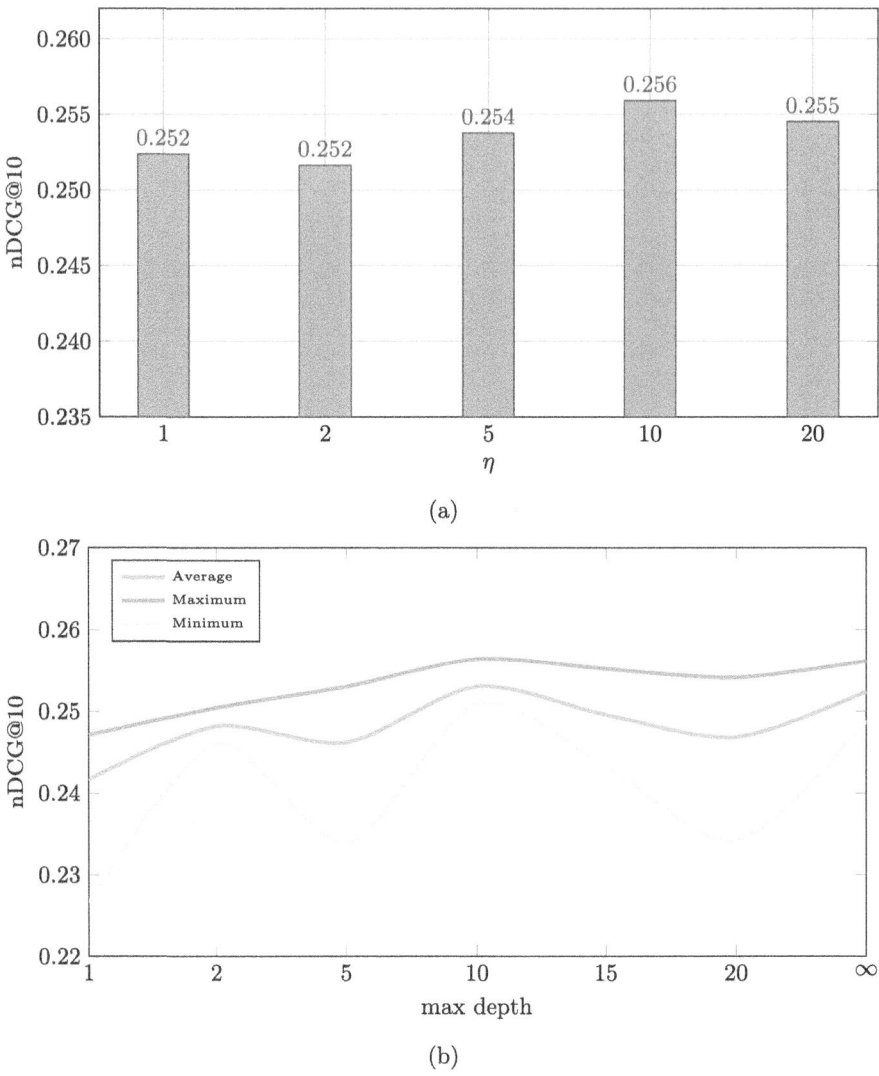

Fig. 3. An example of how KGUF models the item representation. Figure 3a shows the entities in the knowledge graph linked to the items. In Fig. 3b, the most relevant semantic features selected by KGUF are reported.

Effect of Negative Sampling. We investigate how KGUF behaves when we enhance the negative sampling employed in constructing decision trees by introducing a number of negative items equal to η times the number of positive items, such that $|\tilde{\mathcal{I}}_u^-| = \eta \, |\mathcal{I}_u^+|$. The results depicted in Fig. 3a indicate an improvement in accuracy performance with respect to the target metric as η increases. We attribute this trend to the higher values of η mitigating the influence of randomly selecting negative items, consequently reducing noise impact. However, it is important to note that increasing the value of η does not guarantee a constant improvement in performance. Therefore, there is no ideal value for η, and this parameter must be explored individually for each dataset.

Effect of Tree Depth. The maximum depth of a decision tree, representing the length of the longest path from the tree root to a leaf, directly influences the selection of features. We investigate how imposing constraints on the tree depth may impact the recommendation performance. To address this, we trained the model with diverse maximum tree depths-1, 2, 5, 10, 15, 20, and without a limit (∞). We evaluated accuracy performance repeating the experiment five times and presenting the average, minimum, and maximum values of the target metric in Fig. 3b. While unlimited depth yields satisfactory results, it is noteworthy that exploring the entire tree incurs higher computational and memory costs. Excessive features can lead to an overly generalized model, and comparable or even superior performance can be achieved without traversing the entire tree, as it is evident at depth 10, where we witness the least variance around the average performance.

7 Conclusion and Future Work

This paper addresses the need for simplification in existing KGCF recommender systems, proposing a novel algorithm called **KGUF** (Knowledge Graph User-based Filtering). We advocate for a streamlined approach, asserting that simpler strategies have the potential to yield comparable or even superior accuracy.

In KGUF, we focus on efficiently utilizing KG side information by learning user and item embeddings through linear propagation on the user-item interaction graph. Unlike other approaches, our solution integrates latent representations for semantic features directly into item representations. Additionally, we exploit the simple but effective decision tree mechanism to select meaningful semantic features from KGs and filter out noise triples.

We conduct extensive experiments on well-known recommendation datasets, evaluating the effectiveness of our framework. Furthermore, we explore improving the decision tree capability in selecting relevant semantic features. We investigate the impact of negative sampling and tree depth constraints and evaluate their impact on the overall recommendation performance.

References

1. Anelli, V.W., et al.: Elliot: a comprehensive and rigorous framework for reproducible recommender systems evaluation. In: SIGIR, pp. 2405–2414. ACM (2021)
2. Anelli, V.W., et al.: Reshaping graph recommendation with edge graph collaborative filtering and customer reviews. In: DL4SR@CIKM. CEUR Workshop Proceedings, vol. 3317. CEUR-WS.org (2022)

3. Anelli, V.W., et al.: Challenging the myth of graph collaborative filtering: a reasoned and reproducibility-driven analysis. In: RecSys, pp. 350–361. ACM (2023)
4. Anelli, V.W., Noia, T.D., Sciascio, E.D., Ferrara, A., Mancino, A.C.M.: Sparse feature factorization for recommender systems with knowledge graphs. In: RecSys, pp. 154–165. ACM (2021)
5. Anelli, V.W., Di Noia, T., Di Sciascio, E., Ragone, A., Trotta, J.: How to make latent factors interpretable by feeding factorization machines with knowledge graphs. In: Ghidini, C., et al. (eds.) ISWC 2019. LNCS, vol. 11778, pp. 38–56. Springer, Cham (2019). https://doi.org/10.1007/978-3-030-30793-6_3
6. Anelli, V.W., Noia, T.D., Sciascio, E.D., Ragone, A., Trotta, J.: Semantic interpretation of top-n recommendations. IEEE Trans. Knowl. Data Eng. **34**(5), 2416–2428 (2022)
7. Cao, Y., Wang, X., He, X., Hu, Z., Chua, T.: Unifying knowledge graph learning and recommendation: towards a better understanding of user preferences. In: Liu, L., et al. (eds.) The World Wide Web Conference, WWW 2019, San Francisco, CA, USA, 13-17 May 2019, pp. 151–161. ACM (2019). https://doi.org/10.1145/3308558.3313705
8. Chen, L., Wu, L., Hong, R., Zhang, K., Wang, M.: Revisiting graph based collaborative filtering: a linear residual graph convolutional network approach. In: AAAI, pp. 27–34. AAAI Press (2020)
9. Chen, Y., Yang, Y., Wang, Y., Bai, J., Song, X., King, I.: Attentive knowledge-aware graph convolutional networks with collaborative guidance for personalized recommendation. In: ICDE, pp. 299–311. IEEE (2022)
10. Covington, P., Adams, J., Sargin, E.: Deep neural networks for youtube recommendations. In: Sen, S., Geyer, W., Freyne, J., Castells, P. (eds.) Proceedings of the 10th ACM Conference on Recommender Systems, Boston, MA, USA, 15-19 September 2016, pp. 191–198. ACM (2016). https://doi.org/10.1145/2959100.2959190
11. Du, Y., Zhu, X., Chen, L., Zheng, B., Gao, Y.: HAKG: hierarchy-aware knowledge gated network for recommendation. In: SIGIR, pp. 1390–1400. ACM (2022)
12. Ebesu, T., Shen, B., Fang, Y.: Collaborative memory network for recommendation systems. In: Collins-Thompson, K., Mei, Q., Davison, B.D., Liu, Y., Yilmaz, E. (eds.) The 41st International ACM SIGIR Conference on Research & Development in Information Retrieval, SIGIR 2018, Ann Arbor, MI, USA, 08-12 July 2018, pp. 515–524. ACM (2018). https://doi.org/10.1145/3209978.3209991, https://doi.org/10.1145/3209978.3209991
13. Ferrara, A., Anelli, V.W., Mancino, A.C.M., Noia, T.D., Sciascio, E.D.: Kgflex: efficient recommendation with sparse feature factorization and knowledge graphs. ACM Trans. Recomm. Syst. (2023). https://doi.org/10.1145/3588901, just Accepted
14. Gao, B., Liu, T., Wei, W., Wang, T., Li, H.: Semi-supervised ranking on very large graphs with rich metadata. In: Apté, C., Ghosh, J., Smyth, P. (eds.) Proceedings of the 17th ACM SIGKDD International Conference on Knowledge Discovery and Data Mining, San Diego, CA, USA, 21-24 August 2011, pp. 96–104. ACM (2011). https://doi.org/10.1145/2020408.2020430, https://doi.org/10.1145/2020408.2020430
15. de Gemmis, M., Lops, P., Musto, C., Narducci, F., Semeraro, G.: Semantics-aware content-based recommender systems. In: Ricci, F., Rokach, L., Shapira, B. (eds.) Recommender Systems Handbook, pp. 119–159. Springer, Boston, MA (2015). https://doi.org/10.1007/978-1-4899-7637-6_4

16. Glorot, X., Bengio, Y.: Understanding the difficulty of training deep feedforward neural networks. In: AISTATS. JMLR Proceedings, vol. 9, pp. 249–256. JMLR.org (2010)
17. Gunawardana, A., Shani, G.: Evaluating recommender systems. In: Recommender Systems Handbook, pp. 265–308. Springer (2015)
18. Hamilton, W.L.: Graph Representation Learning. Morgan & Claypool Publishers, Synthesis Lectures on Artificial Intelligence and Machine Learning (2020)
19. He, X., Deng, K., Wang, X., Li, Y., Zhang, Y., Wang, M.: Lightgcn: simplifying and powering graph convolution network for recommendation. In: Huang, J.X., Chang, Y., Cheng, X., Kamps, J., Murdock, V., Wen, J., Liu, Y. (eds.) Proceedings of the 43rd International ACM SIGIR conference on research and development in Information Retrieval, SIGIR 2020, Virtual Event, China, 25-30 July 2020, pp. 639–648. ACM (2020). https://doi.org/10.1145/3397271.3401063, https://doi.org/10.1145/3397271.3401063
20. He, X., Liao, L., Zhang, H., Nie, L., Hu, X., Chua, T.: Neural collaborative filtering. In: Barrett, R., Cummings, R., Agichtein, E., Gabrilovich, E. (eds.) Proceedings of the 26th International Conference on World Wide Web, WWW 2017, Perth, Australia, 3-7 April 2017, pp. 173–182. ACM (2017). https://doi.org/10.1145/3038912.3052569
21. Huang, R., Han, C., Cui, L.: Entity-aware collaborative relation network with knowledge graph for recommendation. In: CIKM, pp. 3098–3102. ACM (2021)
22. Kipf, T.N., Welling, M.: Semi-supervised classification with graph convolutional networks. In: ICLR (Poster). OpenReview.net (2017)
23. Koren, Y.: Factor in the neighbors: Scalable and accurate collaborative filtering. ACM Trans. Knowl. Discov. Data 4(1), 1:1–1:24 (2010)
24. Koren, Y., Bell, R.M., Volinsky, C.: Matrix factorization techniques for recommender systems. Computer 42(8), 30–37 (2009)
25. Li, J., Xu, Z., Tang, Y., Zhao, B., Tian, H.: Deep hybrid knowledge graph embedding for Top-N recommendation. In: Wang, G., Lin, X., Hendler, J., Song, W., Xu, Z., Liu, G. (eds.) WISA 2020. LNCS, vol. 12432, pp. 59–70. Springer, Cham (2020). https://doi.org/10.1007/978-3-030-60029-7_6
26. Mancino, A.C.M., Ferrara, A., Bufi, S., Malitesta, D., Noia, T.D., Sciascio, E.D.: Kgtore: tailored recommendations through knowledge-aware GNN models. In: Zhang, J., et al. (eds.) Proceedings of the 17th ACM Conference on Recommender Systems, RecSys 2023, Singapore, Singapore, 18-22 September 2023, pp. 576–587. ACM (2023). https://doi.org/10.1145/3604915.3608804
27. Mao, K., et al.: Simplex: a simple and strong baseline for collaborative filtering. In: CIKM, pp. 1243–1252. ACM (2021)
28. Mao, K., Zhu, J., Xiao, X., Lu, B., Wang, Z., He, X.: Ultragcn: ultra simplification of graph convolutional networks for recommendation. In: CIKM, pp. 1253–1262. ACM (2021)
29. Rendle, S., Freudenthaler, C., Gantner, Z., Schmidt-Thieme, L.: BPR: bayesian personalized ranking from implicit feedback. In: Bilmes, J.A., Ng, A.Y. (eds.) UAI 2009, Proceedings of the Twenty-Fifth Conference on Uncertainty in Artificial Intelligence, Montreal, QC, Canada, 18-21 June 2009, pp. 452–461. AUAI Press (2009), https://www.auai.org/uai2009/papers/UAI2009_0139_48141db02b9f0b02bc7158819ebfa2c7.pdf
30. Shen, Y., et al.: How powerful is graph convolution for recommendation? In: CIKM, pp. 1619–1629. ACM (2021)
31. l Shuai, J., et al.: A review-aware graph contrastive learning framework for recommendation. In: SIGIR, pp. 1283–1293. ACM (2022)

32. Spillo, G., Musto, C., de Gemmis, M., Lops, P., Semeraro, G.: Exploiting neurosymbolic graph embeddings based on first-order logical rules for knowledge-aware recommendations. In: DP@AI*IA. CEUR Workshop Proceedings, vol. 3419, pp. 1–11. CEUR-WS.org (2022)
33. Steck, H.: Evaluation of recommendations: rating-prediction and ranking. In: RecSys, pp. 213–220. ACM (2013)
34. Tian, C., Xie, Y., Li, Y., Yang, N., Zhao, W.X.: Learning to denoise unreliable interactions for graph collaborative filtering. In: SIGIR, pp. 122–132. ACM (2022)
35. Tiddi, I., Lécué, F., Hitzler, P. (eds.): Knowledge Graphs for eXplainable Artificial Intelligence: Foundations, Applications and Challenges, Studies on the Semantic Web, vol. 47. IOS Press (2020)
36. Togashi, R., Otani, M., Satoh, S.: Alleviating cold-start problems in recommendation through pseudo-labelling over knowledge graph. In: WSDM, pp. 931–939. ACM (2021)
37. Velickovic, P., Cucurull, G., Casanova, A., Romero, A., Liò, P., Bengio, Y.: Graph attention networks. In: ICLR (Poster). OpenReview.net (2018)
38. Wang, H., et al.: Exploring high-order user preference on the knowledge graph for recommender systems. ACM Trans. Inf. Syst. **37**(3), 32:1–32:26 (2019)
39. Wang, X., He, X., Cao, Y., Liu, M., Chua, T.: KGAT: knowledge graph attention network for recommendation. In: Teredesai, A., Kumar, V., Li, Y., Rosales, R., Terzi, E., Karypis, G. (eds.) Proceedings of the 25th ACM SIGKDD International Conference on Knowledge Discovery & Data Mining, KDD 2019, Anchorage, AK, USA, 4-8 August 2019, pp. 950–958. ACM (2019). https://doi.org/10.1145/3292500.3330989
40. Wang, X., He, X., Wang, M., Feng, F., Chua, T.: Neural graph collaborative filtering. In: SIGIR, pp. 165–174. ACM (2019)
41. Wang, X., et al.: Learning intents behind interactions with knowledge graph for recommendation. In: Leskovec, J., Grobelnik, M., Najork, M., Tang, J., Zia, L. (eds.) WWW 2021: The Web Conference 2021, Virtual Event / Ljubljana, Slovenia, 19-23 April 2021. pp. 878–887. ACM / IW3C2 (2021). https://doi.org/10.1145/3442381.3450133, https://doi.org/10.1145/3442381.3450133
42. Wang, X., Jin, H., Zhang, A., He, X., Xu, T., Chua, T.: Disentangled graph collaborative filtering. In: Huang, J.X., et al. (eds.) Proceedings of the 43rd International ACM SIGIR conference on research and development in Information Retrieval, SIGIR 2020, Virtual Event, China, 25-30 July 25-30 2020, pp. 1001–1010. ACM (2020). https://doi.org/10.1145/3397271.3401137
43. Wang, Z., Lin, G., Tan, H., Chen, Q., Liu, X.: CKAN: collaborative knowledge-aware attentive network for recommender systems. In: SIGIR, pp. 219–228. ACM (2020)
44. Wu, J., et al.: Self-supervised graph learning for recommendation. In: SIGIR, pp. 726–735. ACM (2021)
45. Yang, Y., Huang, C., Xia, L., Li, C.: Knowledge graph contrastive learning for recommendation. In: SIGIR, pp. 1434–1443. ACM (2022)
46. Ying, R., He, R., Chen, K., Eksombatchai, P., Hamilton, W.L., Leskovec, J.: Graph convolutional neural networks for web-scale recommender systems. In: Guo, Y., Farooq, F. (eds.) Proceedings of the 24th ACM SIGKDD International Conference on Knowledge Discovery & Data Mining, KDD 2018, London, UK, 19-23 August 2018, pp. 974–983. ACM (2018). https://doi.org/10.1145/3219819.3219890
47. Yu, J., Xia, X., Chen, T., Cui, L., Hung, N.Q.V., Yin, H.: Xsimgcl: towards extremely simple graph contrastive learning for recommendation. IEEE Trans. Knowl. Data Eng. **36**(2), 913–926 (2024)

48. Yuan, F., He, X., Karatzoglou, A., Zhang, L.: Parameter-efficient transfer from sequential behaviors for user modeling and recommendation. In: Huang, J.X., Chang, Y., Cheng, X., Kamps, J., Murdock, V., Wen, J., Liu, Y. (eds.) Proceedings of the 43rd International ACM SIGIR conference on research and development in Information Retrieval, SIGIR 2020, Virtual Event, China, 25-30 July 2020, pp. 1469–1478. ACM (2020). https://doi.org/10.1145/3397271.3401156, https://doi.org/10.1145/3397271.3401156
49. Zou, D., et al.: Multi-level cross-view contrastive learning for knowledge-aware recommender system. In: Amigó, E., Castells, P., Gonzalo, J., Carterette, B., Culpepper, J.S., Kazai, G. (eds.) SIGIR '22: The 45th International ACM SIGIR Conference on Research and Development in Information Retrieval, Madrid, Spain, 11 - 15 July 2022, pp. 1358–1368. ACM (2022). https://doi.org/10.1145/3477495.3532025, https://doi.org/10.1145/3477495.3532025
50. Zou, D., et al.: Improving knowledge-aware recommendation with multi-level interactive contrastive learning. In: CIKM, pp. 2817–2826. ACM (2022)

The Effectiveness of Graph Contrastive Learning on Mathematical Information Retrieval

Pei-Syuan Wang and Hung-Hsuan Chen

National Central University, Taoyuan, Taiwan
hhchen1105@acm.org

Abstract. This paper details an empirical investigation into using Graph Contrastive Learning (GCL) to generate mathematical equation representations, a critical aspect of Mathematical Information Retrieval (MIR). Our findings reveal that this simple approach consistently exceeds the performance of the current leading formula retrieval model, TangentCFT. To support ongoing research and development in this field, we have made our source code accessible to the public at https://github.com/WangPeiSyuan/GCL-Formula-Retrieval/.

Keywords: Mathematical information retrieval · Graphical contrastive learning · Layout

1 Introduction

Search engines have revolutionized information access, enabling users to locate relevant textual content from the Internet quickly. Meanwhile, academic search engines and digital libraries, such as Google Scholar, CiteSeerX, and PubMed [3, 22], have become indispensable tools in the academic field, allowing researchers to discover related works from a vast amount of documents. Although a general-purpose search engine and an academic search engine may use different strategies to evaluate the quality of a document (e.g., a search engine may analyze the hyperlink structures to infer the importance of a webpage, while an academic search engine may rely on the citation counts to gauge the quality of a paper [4]), they often rely on similar strategies to define the relevance score between a query term and a document. Popular techniques include term frequency statistics (e.g., TFIDF and its variants [9]) and distributed representations learning (e.g., Word2Vec, fastText, and Transformer [16]).

Mathematical formulas commonly play a central role in scientific papers, facilitating the precise expression of abstract ideas. It is crucial to develop methodologies that can effectively retrieve documents that contain mathematical formulas similar to a target formula. Unfortunately, the search for mathematical formulas

---

We appreciate support from the National Science and Technology Council of Taiwan under grant 110-2222-E-008-005-MY3.

is very different from a regular text-based search. While textual search algorithms focus primarily on word frequencies, syntactic structures, and semantic associations, formula search requires a more profound comprehension of mathematical expressions, their inherent structures, and relationships between mathematical entities. Developing mathematical information retrieval (MIR) algorithms involves two main challenges. First, the model needs to capture the notation structure effectively. In MIR, the notation structure is perhaps more critical than string matching and term frequencies. For example, the quadratic equations $ax^2 + bx + c = 0$ and $\alpha\theta^2 + \beta\theta + \gamma = 0$ may convey the same concept, although they contain very different symbols. However, their structures are identical if we represent both equations using parse trees. Second, the labeled relevance score between pairs of mathematical formulas is needed for supervised training. Unfortunately, such datasets are limited. As a result, it could be difficult to apply machine-learned ranking (a.k.a. learning-to-rank) [13] methodologies. These obstacles make MIR still an extremely challenging task.

In this paper, we experiment with applying graph contrastive learning (GCL) on the graph generated from the formula structure to capture the notation structure without the help of labeled relevance scores between formulas, thus addressing the above two challenges. We define the similarity score between each pair of formulas based on the cosine similarity between their embeddings. We conduct experiments using the NTCIR-12 MathIR Wikipedia Formula Browsing Task [25]. Experimental results show that our model consistently outperforms the TangentCFT model [15], the state-of-the-art model for retrieving mathematical formulas. Note that our study focuses exclusively on models that rely solely on mathematical formulas for MIR. Therefore, models like MathBERT or CocoMAE [17,27], which also incorporate additional information such as contextual texts, fall outside the scope of our analysis.

2 Related Work

This section reviews various mathematics information retrieval methodologies, mainly focusing on TangentCFT, a state-of-the-art method.

2.1 Analyze Formula Using Text

Text-based MIR methods convert mathematical formulas into text formats such as LaTeX and MathML and use text similarity measures to assess the similarity between formulas. An example of this approach is the TF-IDF-based method called Math Indexer and Searcher (MIaS) [18]. It represents formulas in MathML format within an XHTML document and considers text and math formula components. However, this method largely overlooks mathematical formulas' structural and semantic aspects and relies mainly on a textual comparison of words and their frequencies [15].

Other approaches employ complex natural language processing models to handle semantic retrieval. For example, Thanda et al. [21] utilized the PV-

DBOW model to learn the embeddings of text paragraphs. Gao et al. [7] proposed the symbol2vec and formula2vec models, which are based on the Continuous Bag-of-Words (CBOW) and Doc2Vec architectures [12,16], respectively, to learn embeddings. These approaches generally transform formulas into vector representations using semantic representation methods. However, these methods minimally consider the structure of the formulas.

2.2 Analyze Formula Using Graph/tree

Tree-based methods consider the symbol structure and arrangement of mathematical formulas by representing them in a structured format, such as a Symbol Layout Tree (SLT) or an Operator Tree (OPT), and compare the similarity of the structures to perform retrieval. Details of SLT and OPT will be introduced in Sect. 3.1.

Among tree-based methods, some compared the similarity of two tree structures by matching paths from the root node to the child nodes [8,26]. However, a successful match requires complete matching of root-to-leaf paths. Yokoi et al. [23] propose a more flexible method by extracting subpaths from the root node to the child nodes and performing matching based on these subpaths, thus increasing the success rate of matching. The MCAT method [11] used both OPT and SLT for path extraction, incorporating path features and information about the sibling nodes, combined with text-based search, to achieve better results. Another method, Approach0 [28], used only OPT for path extraction, generating paths representing subexpressions of mathematical formulas. The similarity calculation is based on the largest common subexpression among the formulas.

2.3 Integrating both Formulas and Contextual Texts

Some studies leveraged the formulas and contextual texts for math information retrieval. For example, MathBERT [17], motivated by the success of pre-trained language models in natural language processing, utilized math formula, its layout, and the contextual texts into a Transformer for training. Coco-MAE [27] integrated the formula and textual information by contrastive learning. These studies leverage mathematical expressions and contextual texts, often leading to promising results in precision and recall.

2.4 TangentCFT

TangentCFT analyzes a formula based only on the formula but not the contextual text. TangentCFT begins by representing mathematical formulas using OPT and SLT. Next, TangentCFT traverses the tree and converts the paths in the tree into tuple sequences. These paths are then encoded and used to train embeddings using the fastText model [1]. Finally, mathematical formula embeddings are obtained by averaging the embeddings of the tuples in a formula.

To the best of our knowledge, when considering the retrieval of mathematical formulas without leveraging contextual text information, TangentCFT is

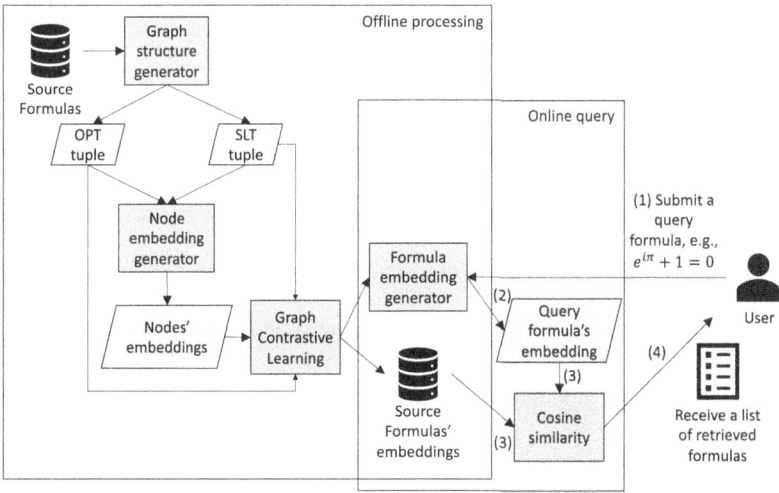

Fig. 1. The online and offline processing of the entire framework

among the effective models currently available [17]. Therefore, this paper uses TangentCFT as the baseline model and compares it with our approach. While it is true that some methods integrate both formulas and contextual texts (as introduced in Sect. 2.3), our study aims to push the boundaries of what can be achieved with formula-only methods before integrating textual data. This focus allows us to isolate the effects and potential of graph-based techniques, contributing to a deeper understanding of their capabilities and limitations.

3 Methodology

We introduce the offline processing module and the online query module in this section. Figure 1 gives an overview of the whole workflow.

3.1 Offline Processing Module

The offline processing module includes a graph structure generator that outputs the OPT and SLT of a formula. We use TangentCFT to generate node (token) embedding, which will be the input of the graph contrastive learning models. The graph contrastive learning models generate formula embeddings based on contrastive learning; thus, the relevance scores between formula pairs are unnecessary. This section details the entire offline processing module.

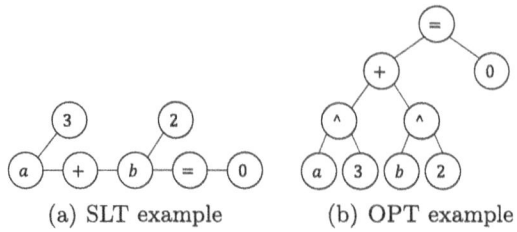

Fig. 2. The examples of the SLT and OPT representations of the formula $a^3 + b^2 = 0$

Table 1. A comparison of the properties of the graph contrastive learning models.

	InfoGraph	GraphCL	BGRL
Requires negative pairs	Y	Y	N
Requires graph augmentation	N	Y	Y
Contrastive pairs	Graph to node	Graph to graph	Node to node
Number of encoders	1	1	2

Graph Structure Generator. A mathematical symbol sequence can form graphs expressing semantic relationships between symbols. This study employs two graph structures to represent the relationship of the symbols in a mathematical formula: Symbol Layout Tree (SLT) and Operator Tree (OPT) [6]. The SLT is used primarily to indicate the spatial positioning of mathematical symbols in a written form. The OPT, on the other hand, is mainly used to capture the semantics of mathematical formulas. The OPT represents operators by an intermediate node, and the child nodes represent operands. Through the commutativity or associativity of operators, mathematically equivalent formulas with different appearances exhibit the same OPT structure.

For example, given a formula $a^3 + b^2 = 0$, Fig. 2 gives its SLT and OPT representations: SLT generates a graph that better preserves the layout of the writing, whereas the output of OPT captures the semantics of the equation.

Token Embedding Generator. We may define the features for the nodes and edges of the SLT and OPT. For example, we could define a feature for a node that specifies whether the token represents an operator or operand. However, manually defining features can be tedious and perhaps subjective. Eventually, we decided to take advantage of the node and edge characteristics described in TangentCFT [15] and apply fastText [1] to the paths sampled by random walks to generate the embeddings for the nodes. We set each output embedding length to 100. These node embeddings are the building blocks for graph embeddings, which are representations of the formulas, as described below.

Formula Embedding Generator and Graph Contrastive Learning. To generate formula embeddings, we need to assemble the node embeddings. There are at least two different ways to do it. The first is to compute the elementwise average for each node embedding in a graph, as TangentCFT does [15]. However, the simple average may be too naïve because the relationship among the math symbols (i.e., nodes) is missing. Another possibility is using the downstream task label as the ground truth and applying backpropagation to learn how to integrate the token embeddings. Unfortunately, our task has a limited number of relevance scores between pairs of formulas. Therefore, learning to integrate token embeddings based on a few labels will likely overfit the training data.

Eventually, we decided to employ graph contrastive learning methods to learn the embeddings of the formulas. GCL generates positive and negative graph pairs by manipulating the graph structures. Thus, the relevance score is not needed during training. We experimented with three representative GCL models: InfoGraph [19], GraphCL [24], and Bootstrapped Graph Latents (BGRL) [20]. Since these models require no training labels, we can generate the formula embedding even if a formula does not appear in the training data.

The InfoGraph model processes multiple graphs in one batch. InfoGraph learns to generate the local node embeddings and the global graph embeddings simultaneously such that a node n_i and a graph g_j have high mutual information if $n_i \in g_j$ and low mutual information otherwise. An advantage of InfoGraph is that it does not rely on graph augmentation techniques. However, InfoGraph assumes that a node's embedding alone can discriminate its belonging graph and other graphs, which could be an over-strong assumption.

The GraphCL model generates a positive graph pair by augmenting a given graph based on, for example, node dropping and edge perturbation. GraphCL regards a negative graph pair by the augmented graphs of two distinct graphs. The loss function encourages positive graph pairs to have similar embeddings and negative graph pairs to have dissimilar embeddings. Although such a technique works exceptionally well in computer vision [5], the data augmentation techniques used in graphs may merit further discussion. For example, in image classification, an image after standard augmenting procedures (e.g., rotating or resizing) would still likely be regarded as having the same label. However, standard graph-augmentation techniques make the graphs structurally different, especially when a graph is small. As a result, the performance of GraphCL may be substantially influenced by the graph-augmenting procedure.

Finally, the BGRL requires only positive pairs generated by graph augmentation. BGRL alleviates the need for negative pairs by applying two distinct encoders: one's parameters are learned via direct backpropagation, and the other's parameters are updated by an exponential moving average of the parameters in the first encoder. Although BGRL is highly scalable because it requires no negative pairs, BGRL still needs graph augmentation, which could still be an issue, as discussed above.

Table 1 compares these popular GCL models. Since each has its strengths and weaknesses, we tested all of them as formula embedding generation methods.

3.2 Online Query Module

A user submits to the system a query formula, which is used by the online query module to generate the query embedding based on the formula embedding generator trained offline. The system computes the cosine similarity between the query formula's embedding and each source formula's embedding. Finally, the system returns a list of the matched formulas by ranking the cosine similarities in descending order.

4 Experiments

4.1 Experimental Dataset and Evaluation Metrics

We used the NTCIR-12 MathIR Wikipedia Formula Browsing Task [25] as the data source for evaluation. Each relevance score is an integer between 0 and 4. We evaluated the results based on binary scores (binary preference, bpref) and graded relevance scores (normalized discounted cumulative gain, nDCG).

The bpref score evaluates a binary retrieval task (relevant/irrelevant) with incomplete information, i.e., the relevance scores of some documents can be unlabeled [2]. The bpref is a perfect evaluation score in our case because the relevance scores between most pairs of documents are unlabeled in our dataset (we have only 1,202 labeled relevance scores). We consider a pair of documents relevant if their relevance score is 3 or 4; otherwise, they are irrelevant.

The definition of bpref is given in Eq. 1.

$$s_{\text{bpref}} = \frac{1}{R} \sum_r \left(1 - \frac{|n \text{ is ranked higher than } r|}{\min(R, N)} \right), \quad (1)$$

where R and N represent the counts of relevant and irrelevant documents, respectively, with r as a relevant and n as an irrelevant document.

Although a binary judgment (relevant/irrelevant) is probably more straightforward for human evaluation [10], it fails to capture a fine-grained assessment. Therefore, we also applied the nDCG evaluation metric because it allows for a graded relevance score. Equation 2 shows the formula of the nDCG score, which is the DCG score normalized by the ideal DCG score.

$$s_{\text{nDCG}} = \frac{s_{\text{DCG}}}{s_{\text{IDCG}}}, \quad (2)$$

where s_{IDCG} is the score of s_{DCG} when the top-K documents are perfectly ordered (i.e., they are ordered according to the relevance score in descending order). The DCG score is computed by Eq. 3.

$$s_{\text{DCG}} = \sum_{i=1}^{K} \frac{r_i}{\log_2(i+1)}, \quad (3)$$

where r_i denotes the ith document's relevance score in the list ($r_i \in \{0, 1, 2, 3, 4\}$), and K is the count of returned documents ($K = 1,000$ in this experiment).

Table 2. The bpref scores of applying different models on SLT layout, OPT layout, and F1 score of the above two. TangentCFT is the baseline; InfoGraph, BGRL, and GraphCL are GCL models used by our approach.

Note	Model	SLT	OPT	F1
Compared baseline	TangentCFT	0.680 ± 0.0053	0.660 ± 0.0064	0.670
GCLs used by the paper	InfoGraph	0.691 ± 0.0066	0.685 ± 0.0070	0.688
	BGRL	$\mathbf{0.701 \pm 0.0089}$	0.683 ± 0.0077	0.692
	GraphCL	0.685 ± 0.0090	$\mathbf{0.703 \pm 0.0072}$	**0.694**

Table 3. The nDCG scores of applying different models on SLT layout, OPT layout, and F1 score of the above two. TangentCFT is the baseline; InfoGraph, BGRL, and GraphCL are GCL models used by our approach.

Note	Model	SLT	OPT	F1
Compared baseline	TangentCFT	0.841 ± 0.0032	0.830 ± 0.0041	0.835
GCLs used by the paper	InfoGraph	$\mathbf{0.860 \pm 0.0036}$	0.851 ± 0.0063	0.855
	BGRL	0.851 ± 0.0075	0.827 ± 0.0078	0.839
	GraphCL	0.855 ± 0.0029	$\mathbf{0.864 \pm 0.0065}$	**0.859**

The nDCG is valid only when all returned documents are scored for relevance. We filter out unjudged formulas from the list. Given each query has at most 90 judged formulas, and our $K = 1,000$ exceeds this, we utilize all judged documents.

In general, the nDCG score measures the effectiveness of a ranking algorithm by considering the relevance and position of items in a ranked list. It places a higher emphasis on the top positions. Additionally, nDCG accommodates graded relevance judgments, allowing for finer distinctions in the relevance of items. Meanwhile, the bpref allows unjudged documents in the list, and the binary judgment is likely more intuitive for most evaluators. Since the two metrics assess the quality of ranked lists from different perspectives, we use both for evaluations.

4.2 Quantitative Result

Table 2 and Table 3 present the quantitative evaluation results of the bpref and nDCG scores when applying different GCL models on the SLT or OPT layouts. We repeat each experiment 5 times and report the mean and standard deviation in these tables. We also report the F1 score for the mean of SLT and the mean of OPT scores. Both the bpref and the nDCG metrics indicate that our self-supervised graph contrastive learning consistently achieves better retrieval performance than TangentCFT, and the results are very stable (since the standard deviations are close to 0). In particular, the bpref score implies that, on average, the genuinely relevant formulas retrieved by our model rank higher than irrelevant ones more often when compared to the state-of-the-art TangentCFT. The nDCG scores also indicate that our method is better at ranking the most relevant formulas near the top.

Interestingly, various GCLs are more effective with different layouts: GraphCL works better when OPT is used, while InfoGraph and BGRL are more successful when using SLT. We show the F1 score in the last columns of Table 2 and Table 3 to show the average effectiveness of each model on different layouts.

4.3 Case Study

Table 4. The top returns of various models when querying "$O(mn \log m)$" (using SLT as the layout for graph construction.)

Rank	InfoGraph	GraphCL	BGRL
1	$O(mn \log m)$	$O(mn \log m)$	$O(mn \log m)$
2	$O(VE \log V)$	$O(n \log m)$	$O(m \log n)$
3	$O(nk \log k)$	$O(m \log n)$	$O(n \log m)$
4	$O(KN \log N)$	$O(mn)$	$O(mn)$
5	$O(m + \log n)$	$O(m^n)$	$O(m^n)$

Table 5. The top returns of various models when querying "$O(mn \log m)$" (using OPT as the layout for graph construction.)

Rank	InfoGraph	GraphCL	BGRL
1	$O(mn \log m)$	$O(mn \log m)$	$O(mn \log m)$
2	$O(n \log m)$	$O(n \log m)$	$O(mn)$
3	$O(m \log n)$	$O(mn \log(mn))$	$O(Mr)$
4	$O(n \log k)$	$O(m^2 n \log n)$	$O(mnp)$
5	$O(mn)$	$O(m \log n \log \log n)$	$\Theta(mn)$

This section shows the top retrieved formulas for two highly distinct query formulas. The first query involves a big-O expression with a logarithmic operation. Big-O notation is standard in analyzing algorithms' time complexity. The inclusion of the log operation introduces mathematical complexity. Retrieving relevant results for such queries evaluates a model's capacity to deal with logarithmic functions, multiplications, and the big-O notation. The second query is an equation represented in matrix form. Equations involving matrices are prevalent in various scientific and engineering fields, including linear algebra, physics,

Table 6. The top returns of various models when querying "$\begin{bmatrix} V_1 \\ I_2 \end{bmatrix} = \begin{bmatrix} h_{11} & h_{12} \\ h_{21} & h_{22} \end{bmatrix} \begin{bmatrix} I_1 \\ V_2 \end{bmatrix}$," (using SLT as the layout for graph construction.)

Rank	InfoGraph	GraphCL	BGRL
1	$\begin{bmatrix} V_1 \\ I_2 \end{bmatrix} = \begin{bmatrix} h_{11} & h_{12} \\ h_{21} & h_{22} \end{bmatrix} \begin{bmatrix} I_1 \\ V_2 \end{bmatrix}$	$\begin{bmatrix} V_1 \\ I_2 \end{bmatrix} = \begin{bmatrix} h_{11} & h_{12} \\ h_{21} & h_{22} \end{bmatrix} \begin{bmatrix} I_1 \\ V_2 \end{bmatrix}$	$\begin{bmatrix} V_1 \\ I_2 \end{bmatrix} = \begin{bmatrix} h_{11} & h_{12} \\ h_{21} & h_{22} \end{bmatrix} \begin{bmatrix} I_1 \\ V_2 \end{bmatrix}$
2	$\begin{bmatrix} V_1 \\ I_1 \end{bmatrix} = \begin{bmatrix} A & B \\ C & D \end{bmatrix} \begin{bmatrix} V_2 \\ -I_2 \end{bmatrix}$	$\begin{bmatrix} h_{11} & h_{12} \\ h_{21} & h_{22} \end{bmatrix}$	$\begin{bmatrix} I_1 \\ V_2 \end{bmatrix} = \begin{bmatrix} g_{11} & g_{12} \\ g_{21} & g_{22} \end{bmatrix} \begin{bmatrix} V_1 \\ I_2 \end{bmatrix}$
3	$\begin{bmatrix} I_1 \\ V_2 \end{bmatrix} = \begin{bmatrix} g_{11} & g_{12} \\ g_{21} & g_{22} \end{bmatrix} \begin{bmatrix} V_1 \\ I_2 \end{bmatrix}$	$s_{(3,2,2,1)} = \begin{vmatrix} h_3 & h_4 & h_5 & h_6 \\ h_1 & h_2 & h_3 & h_4 \\ 1 & h_1 & h_2 & h_3 \\ 0 & 0 & 1 & h_1 \end{vmatrix}$	$\begin{bmatrix} V_1 \\ I_1 \end{bmatrix} = \begin{bmatrix} A & B \\ C & D \end{bmatrix} \begin{bmatrix} V_2 \\ -I_2 \end{bmatrix}$
4	$\begin{bmatrix} V_1 \\ V_2 \end{bmatrix} = \begin{bmatrix} 0 & -r \\ r & 0 \end{bmatrix} \begin{bmatrix} I_1 \\ I_2 \end{bmatrix}$	$\begin{bmatrix} \frac{1}{h_{11}} & \frac{-h_{12}}{h_{11}} \\ \frac{h_{21}}{h_{11}} & \frac{\Delta[h]}{h_{11}} \end{bmatrix}$	$\begin{bmatrix} V_2 \\ I'_2 \end{bmatrix} = \begin{bmatrix} 1 & -R \\ -sC & 1+sCR \end{bmatrix} \begin{bmatrix} V_1 \\ I_1 \end{bmatrix}$
5	$\begin{bmatrix} h_{11} & h_{12} \\ h_{21} & h_{22} \end{bmatrix}$	$h_{11} = \frac{V_1}{I_1}\big|_{V_2=0}$	$\begin{bmatrix} h_{11} & h_{12} \\ h_{21} & h_{22} \end{bmatrix}$

and computer graphics. Retrieving relevant results for matrix equations is essential in applications like solving linear systems or optimizing operations on large datasets.

Tables 4 and 5 show the top-5 returns of the GCL models for the query with the big-O and logarithm expression. All the best-matched formulas are precisely the query formula. Additionally, all returns involve the big-O notation, except the 5th return of BGRL using OPT, which retrieves a highly relevant big-Θ notation. Also, some formulas with semantics identical to "$O(mn \log m)$" but using different symbols, such as $O(VE \log V)$ or $O(KN \log N)$, are retrieved, indicating that these models effectively handles polynomials, logarithm, and the big-O notation.

Table 6 and Table 7 show the top-5 formulas retrieved from the query with the matrix equation. We arrive at the same findings as in the previous case. First, all the top returns are the same as in the query formula. Moreover, the top-5 returns of the models using SLT and OPT as the layout for graph construction all contain matrices, except the 5th return of GraphCL using SLT. In addition, models can retrieve semantically similar formulas with different symbols.

5 Discussion

In this study, we investigate graph contrastive learning for formula retrieval to address two challenges of mathematical information retrieval: the model needs to capture the notation structure and the lack of relevance score between formula pairs. We explore the potential of popular GCL methods, including InfoGraph,

Table 7. The top returns of various models when querying "$\begin{bmatrix} V_1 \\ I_2 \end{bmatrix} = \begin{bmatrix} h_{11} & h_{12} \\ h_{21} & h_{22} \end{bmatrix} \begin{bmatrix} I_1 \\ V_2 \end{bmatrix}$," (using OPT as the layout for graph construction.)

Rank	InfoGraph	GraphCL	BGRL
1	$\begin{bmatrix} V_1 \\ I_2 \end{bmatrix} = \begin{bmatrix} h_{11} & h_{12} \\ h_{21} & h_{22} \end{bmatrix} \begin{bmatrix} I_1 \\ V_2 \end{bmatrix}$	$\begin{bmatrix} V_1 \\ I_2 \end{bmatrix} = \begin{bmatrix} h_{11} & h_{12} \\ h_{21} & h_{22} \end{bmatrix} \begin{bmatrix} I_1 \\ V_2 \end{bmatrix}$	$\begin{bmatrix} V_1 \\ I_2 \end{bmatrix} = \begin{bmatrix} h_{11} & h_{12} \\ h_{21} & h_{22} \end{bmatrix} \begin{bmatrix} I_1 \\ V_2 \end{bmatrix}$
2	$\begin{bmatrix} I_1 \\ V_2 \end{bmatrix} = \begin{bmatrix} g_{11} & g_{12} \\ g_{21} & g_{22} \end{bmatrix} \begin{bmatrix} V_1 \\ I_2 \end{bmatrix}$	$\begin{pmatrix} I_1 \\ I_2 \end{pmatrix} = \begin{pmatrix} Y_{11} & Y_{12} \\ Y_{21} & Y_{22} \end{pmatrix} \begin{pmatrix} V_1 \\ V_2 \end{pmatrix}$	$\begin{bmatrix} I_1 \\ V_2 \end{bmatrix} = \begin{bmatrix} g_{11} & g_{12} \\ g_{21} & g_{22} \end{bmatrix} \begin{bmatrix} V_1 \\ I_2 \end{bmatrix}$
3	$\begin{bmatrix} V_1 \\ V_2 \end{bmatrix} = \begin{bmatrix} z_{11} & z_{12} \\ z_{21} & z_{22} \end{bmatrix} \begin{bmatrix} I_1 \\ I_2 \end{bmatrix}$	$\begin{bmatrix} \frac{1}{h_{11}} & \frac{-h_{12}}{h_{11}} \\ \frac{h_{21}}{h_{11}} & \frac{\Delta[\mathbf{h}]}{h_{11}} \end{bmatrix}$	$\begin{bmatrix} I_1 \\ I_2 \end{bmatrix} = \begin{bmatrix} y_{11} & y_{12} \\ y_{21} & y_{22} \end{bmatrix} \begin{bmatrix} V_1 \\ V_2 \end{bmatrix}$
4	$\begin{pmatrix} a_1 \\ b_1 \end{pmatrix} = \begin{pmatrix} T_{11} & T_{12} \\ T_{21} & T_{22} \end{pmatrix} \begin{pmatrix} b_2 \\ a_2 \end{pmatrix}$	$\begin{bmatrix} h_{11} & h_{12} \\ h_{21} & h_{22} \end{bmatrix}$	$\begin{bmatrix} K_{11} & K_{12} \\ K_{21} & K_{22} \end{bmatrix} \begin{bmatrix} x_1 \\ x_2 \end{bmatrix} = \begin{bmatrix} F_1 \\ F_2 \end{bmatrix}$
5	$\begin{bmatrix} K_{11} & K_{12} \\ K_{21} & K_{22} \end{bmatrix} \begin{bmatrix} x_1 \\ x_2 \end{bmatrix} = \begin{bmatrix} F_1 \\ F_2 \end{bmatrix}$	$\begin{bmatrix} \frac{\Delta[\mathbf{h}]}{h_{22}} & \frac{h_{12}}{h_{22}} \\ \frac{-h_{21}}{h_{22}} & \frac{1}{h_{22}} \end{bmatrix}$	$\begin{pmatrix} A_1 & B_1 \\ A_2 & B_2 \end{pmatrix} \begin{pmatrix} x \\ y \end{pmatrix} = \begin{pmatrix} C_1 \\ C_2 \end{pmatrix}$.

GraphCL, and BGRL. We investigate the OPT and SLT graph layouts and their influence on the retrieval results. We observe that the GCL models outperform TangentCFT, a state-of-the-art formula retrieval model. However, TangentCFT is still essential, as our GCL models utilize the node embeddings generated by TangentCFT as the input for these GCL models. We also use case studies to confirm that the methods can handle different formula queries.

To enrich the training instances and further enhance model robustness, future work could explore the generation of new equations as positive training pairs based on equation templates. For example, given a regular expression that generates polynomial equations, each pair of these generated formulas could be a potential positive pair. Another future work could be to improve the GCL data-augmentation process. Current strategies, randomly adding or removing nodes/edges from graphs generated from equations, may not always be ideal. Such adjustments may alter the semantics of the equation and introduce structural inconsistencies. Thus, we are also interested in developing more sophisticated graph-augmentation strategies that preserve the meaning and structure of the equation while increasing the diversity of training data. Finally, we would also like to experiment with the model on other datasets, e.g., the ARQMath dataset [14].

References

1. Bojanowski, P., Grave, E., Joulin, A., Mikolov, T.: Enriching word vectors with subword information. Trans. Assoc. Comput. Linguist. **5**, 135–146 (2017)
2. Buckley, C., Voorhees, E.M.: Retrieval evaluation with incomplete information. In: Proceedings of the 27th annual international ACM SIGIR Conference on Research and Development in Information Retrieval, pp. 25–32 (2004)
3. Caragea, C., et al.: CiteSeer^x: a scholarly big dataset. In: de Rijke, M., et al. Advances in Information Retrieval, ECIR 2014, LNCS, vol. 8416, pp. 311–322. Springer, Cham (2014). https://doi.org/10.1007/978-3-319-06028-6_26
4. Chen, H.H., Treeratpituk, P., Mitra, P., Giles, C.L.: CSSeer: an expert recommendation system based on CiteseerX. In: Proceedings of the 13th ACM/IEEE-CS Joint Conference on Digital Libraries, pp. 381–382 (2013)
5. Chen, T., Kornblith, S., Norouzi, M., Hinton, G.: A simple framework for contrastive learning of visual representations. In: International Conference on Machine Learning, pp. 1597–1607. PMLR (2020)
6. Davila, K., Zanibbi, R.: Layout and semantics: combining representations for mathematical formula search. In: Proceedings of the 40th International ACM SIGIR Conference on Research and Development in Information Retrieval, pp. 1165–1168 (2017)
7. Gao, L., Jiang, Z., Yin, Y., Yuan, K., Yan, Z., Tang, Z.: Preliminary exploration of formula embedding for mathematical information retrieval: can mathematical formulae be embedded like a natural language? arXiv preprint arXiv:1707.05154 (2017)
8. Hijikata, Y., Hashimoto, H., Nishida, S.: An investigation of index formats for the search of mathml objects. In: 2007 IEEE/WIC/ACM International Conferences on Web Intelligence and Intelligent Agent Technology-Workshops, pp. 244–248. IEEE (2007)
9. Hsu, L.Y., Kao, C.H., Jheng, I.S., Chen, H.H.: Toward building an academic search engine understanding the purposes of the matched sentences in an abstract. IEEE Access **9**, 109344–109354 (2021)
10. Kekäläinen, J.: Binary and graded relevance in IR evaluations-comparison of the effects on ranking of IR systems. Inf. Process. Manage. **41**(5), 1019–1033 (2005)
11. Kristianto, G.Y., Topic, G., Aizawa, A.: MCAT math retrieval system for NTCIR-12 MathIR Task. In: NTCIR (2016)
12. Le, Q., Mikolov, T.: Distributed representations of sentences and documents. In: International Conference on Machine Learning, pp. 1188–1196. PMLR (2014)
13. Liu, T.Y., et al.: Learning to rank for information retrieval. Found. Trends® Inf. Retrieval **3**(3), 225–331 (2009)
14. Mansouri, B., Agarwal, A., Oard, D., Zanibbi, R.: Finding old answers to new math questions: the ARQMath lab at CLEF 2020. In: Jose, J.M., Yilmaz, E., Magalhães, J., Castells, P., Ferro, N., Silva, M.J., Martins, F. (eds.) ECIR 2020, Part II. LNCS, vol. 12036, pp. 564–571. Springer, Cham (2020). https://doi.org/10.1007/978-3-030-45442-5_73
15. Mansouri, B., Rohatgi, S., Oard, D.W., Wu, J., Giles, C.L., Zanibbi, R.: Tangent-CFT: an embedding model for mathematical formulas. In: Proceedings of the 2019 ACM SIGIR International Conference on Theory of Information Retrieval, pp. 11–18 (2019)
16. Mikolov, T., Sutskever, I., Chen, K., Corrado, G.S., Dean, J.: Distributed representations of words and phrases and their compositionality. In: Advances in Neural Information Processing Systems, vol. 26 (2013)

17. Peng, S., Yuan, K., Gao, L., Tang, Z.: Mathbert: a pre-trained model for mathematical formula understanding. arXiv preprint arXiv:2105.00377 (2021)
18. Sojka, P., Líška, M.: The art of mathematics retrieval. In: Proceedings of the 11th ACM Symposium on Document Engineering, pp. 57–60 (2011)
19. Sun, F.Y., Hoffmann, J., Verma, V., Tang, J.: Infograph: unsupervised and semi-supervised graph-level representation learning via mutual information maximization. arXiv preprint arXiv:1908.01000 (2019)
20. Thakoor, S., et al.: Large-scale representation learning on graphs via bootstrapping. arXiv preprint arXiv:2102.06514 (2021)
21. Thanda, A., Agarwal, A., Singla, K., Prakash, A., Gupta, A.: A document retrieval system for math queries. In: NTCIR (2016)
22. Wu, J., et al.: Citeseerx: AI in a digital library search engine. AI Mag. **36**(3), 35–48 (2015)
23. Yokoi, K., Aizawa, A.: An approach to similarity search for mathematical expressions using mathml. Towards a Digital Mathematics Library. Grand Bend, Ontario, Canada, July 8-9th, 2009 pp. 27–35 (2009)
24. You, Y., Chen, T., Sui, Y., Chen, T., Wang, Z., Shen, Y.: Graph contrastive learning with augmentations. Adv. Neural. Inf. Process. Syst. **33**, 5812–5823 (2020)
25. Zanibbi, R., Aizawa, A., Kohlhase, M., Ounis, I., Topic, G., Davila, K.: Ntcir-12 mathir task overview. In: NTCIR (2016)
26. Zhong, W., Fang, H.: OPMES: a similarity search engine for mathematical content. In: Ferro, N., Crestani, F., Moens, M.-F., Mothe, J., Silvestri, F., Di Nunzio, G.M., Hauff, C., Silvello, G. (eds.) ECIR 2016. LNCS, vol. 9626, pp. 849–852. Springer, Cham (2016). https://doi.org/10.1007/978-3-319-30671-1_79
27. Zhong, W., Lin, S.C., Yang, J.H., Lin, J.: One blade for one purpose: advancing math information retrieval using hybrid search. In: Proceedings of the 46th International ACM SIGIR Conference on Research and Development in Information Retrieval (2023)
28. Zhong, W., Zanibbi, R.: Structural similarity search for formulas using leaf-root paths in operator subtrees. In: Azzopardi, L., Stein, B., Fuhr, N., Mayr, P., Hauff, C., Hiemstra, D. (eds.) ECIR 2019, Part I. LNCS, vol. 11437, pp. 116–129. Springer, Cham (2019). https://doi.org/10.1007/978-3-030-15712-8_8

The Impact of Source-Target Node Distance on Vicious Adversarial Attacks in Social Network Recommendation Systems

Federico Albanese[1], Giovanni Trappolini[2](✉), Lorenzo Scarlino[2], and Fabrizio Silvestri[2]

[1] University of Buenos Aires, Buenos Aires, Argentina
[2] Sapienza University, Rome, Italy
{trappolini,scarlino,fsilvestri}@diag.uniroma1.it

Abstract. Social network recommendation systems play a crucial role in shaping user experiences by using graph neural networks and link prediction methods to tailor suggestions for new contacts. However, these systems are vulnerable to adversarial attacks orchestrated by malicious users using frameworks to manipulate recommendations artificially. In particular, frameworks such as SAVAGE have proven that conducting efficient and effective *sparse* vicious attacks on recommendation systems is possible. However, whether these adversarial systems remain effective and how much are impacted by the distance between the source and target nodes in the social network is unclear. This study investigates the impact of source-target path length on overall attack performance, revealing real-world applications such as social media friend/follower suggestions, where, for example, the source user may or may not share common connections with the target user and/or the friends of the target user. Analyzing a Twitter dataset, we found that SAVAGE outperforms other models, especially for longer source-target distances, with minimal impact on the network structure. These results make it highly relevant for real-world scenarios.

Keywords: Graph Neural Network · Adversarial Attack · Link Prediction · Social Media

1 Introduction

In social networks, user interactions and content recommendations are pivotal in shaping user experiences. Graph neural networks and link prediction methods allow for the personalized suggestion of new contacts within a social network [35]. Based on estimated probabilities for a given user, the ranked list of predicted links represents a set of potential new contacts. This becomes particularly crucial

---

F. Albanese and G. Trappolini—Equal contributions.

in the context of social media influencers, authoritative figures with a substantial number of followers. Being followed by such influencers can be construed as a valuable endorsement, leading to various benefits for the follower, such as increased revenue from advertising campaigns.

However, these recommendation systems are susceptible to adversarial attacks, especially by malicious users aiming to boost their reputation on social media platforms artificially [14,38,59]. SAVAGE (SpArse Vicious Attacks on Graph nEtworks) is an attack model that uses vicious nodes to exploit weaknesses in graph neural network-based link prediction systems [50]. Previous works focused on characterizing the model performance with diverse network datasets using random pairs of source-target nodes. However, the success rate of an adversarial attack depends on the distance in the network between the source and the target node. Orchestrating a successful attack might be more challenging when the source node is distant from the target node. To the best of our knowledge, there has been no comprehensive examination of adversarial link-prediction attacks across different source-target node distances.

In this work, we present a comprehensive analysis of SAVAGE performance and its advantages over alternative link prediction adversarial attack methods in terms of attack success and resource optimization. Using a social media dataset composed of tweets during US President Biden's inauguration day, we study the impact of the source-target path length on the overall attack performance, which has real-world applications such as social media friend/follower suggestions where the source user might or might not have friends or friends of friends in common with the target user. The study provides a deeper understanding of how significantly this factor can influence the attack's success by exploring different scenarios where source and target nodes are located at a variable distance in the graph.

The main contributions of this work are:

– **SAVAGE White-Box Adversarial Attack**: a practical SAVAGE attack on a link prediction model, evaluated on the Twitter retweet graph using distance-based sample sets. Results show a 25% performance improvement when comparing before and after the attack, with minimal impact on the network structure.
– **Comparative Adversarial Attack Analysis**: SAVAGE outperforms Iterative Gradient Attack and random baselines in final prediction and attack rate across all distances. SAVAGE's performance is notably better for long distances in comparison with other models, making it highly relevant for real-world scenarios.

The paper contributes novel insights into the effectiveness, impact, and resource utilization of the SAVAGE white-box adversarial attack in link prediction models, establishing its superiority over existing state-of-the-art methods. The comprehensive analysis, robustness assessment, and comparative study with other attacks enhance the understanding of adversarial threats in network-based systems.

2 Related Work

2.1 Link Prediction on Graphs

Recent advancements in deep learning have generated a spur in research activity. Large Language Models (LLMs) [13,30,49] have redefined standards for textual generative tasks. There has also been an effort to make this adapt to other languages [7,10,21,24,32]. But the effort has not been limited to text, with applications ranging from images/videos [27,28,51], music [2,9,20], and many domains and tasks [6,34,46,56]. Of course, relational data, or graphs, and its related problems, has been central in this endeavor.

Particularly, the problem of predicting connections, commonly known as link prediction, is a highly explored area in contemporary graph analysis. For an extensive overview of this topic, the reader is directed to several key references [35,41,55]. A fundamental yet effective group of methods for link prediction is the heuristic techniques. These methods rely on the hypothesis that similar nodes are more likely to be connected than dissimilar nodes. In practice, these techniques employ predetermined heuristics to estimate node similarity scores, which then predict the likelihood of a link [37,39]. These heuristic approaches are classified according to the extent of neighbors considered for computing node similarity scores. For instance, methodologies like Common Neighbors [42], Jaccard [37], and Preferential Attachment [8] fall under first-order heuristics as they only take into account the immediate neighbors of the nodes in question. Conversely, Adamic-Adar [1] and resource allocation strategies [58] are second-order heuristics involving a two-hop neighborhood. Other more complex heuristics, known as high-order heuristics, necessitate knowledge of the entire network, with examples including the Katz index [33], rooted PageRank [12], and SimRank [29]. Despite their practical effectiveness, heuristic methods are often constrained by their inherent assumptions about link existence. To address these limitations, link prediction has been reframed as a typical binary classification problem, tackled using established supervised learning approaches [3]. With the progression of deep learning and graph neural networks (GNNs) in particular, various recent studies have introduced effective GNN-based methods for link prediction [36,57]. Broadly, GNNs facilitate learning appropriate node representations (embeddings) by aggregating data from each node's neighbors. These embeddings are then utilized in a subsequent link prediction function, enabling the entire model architecture to be trained in an end-to-end manner. This paper specifically examines a particular adversarial attack targeting GNN-based link prediction methodologies.

2.2 Adversarial Attacks to Link Prediction

Despite their effectiveness, research indicates that machine learning models are susceptible to what are known as "adversarial attacks." These are intentionally designed, harmful examples aimed at deceiving the models' predictive capabilities. Such adversarial inputs are typically created by applying slight yet carefully chosen changes to standard inputs. These tactics have been proven to be effective in various important areas, like image analysis [25] and malware identification [26]. Additionally, techniques used in adversarial perturbations share features with the sub-field of counterfactual examples [16–19,40,47,48]. However, only a limited number of studies have investigated the effectiveness of these adversarial attacks on link prediction models, particularly those that use Graph Neural Networks (GNNs), which necessitate altering the input graph. Noteworthy among these studies are [14,15,38,59], and a detailed survey can be found in [22]. The mentioned methods generally assume that the attacker has control over a portion of the existing nodes. However, this assumption is often impractical due to the excessive costs involved. To counter this, [54] introduced a method allowing attackers to create new (i.e., counterfeit or malicious) nodes for more efficient attacks. Additionally, [53] developed a linear approximation of this approach for better scalability, and [23] designed a universal perturbation capable of targeting multiple nodes while remaining effective. In both traditional and malicious scenarios, all these methods restrict the extent of the graph modification. This limit is typically enforced in two ways: either by assigning the attacker a specified budget [14,45] or by setting rules to limit the variance between the original and altered graph, as seen in [38]. In either case, this serves as a cap on the amount of "malicious resources" that can be utilized for the attack and does not encourage moderation. For this reason, SAVAGE [50] was proposed. Savage, thanks to a set of purposely crafted losses, manages to mount vicious attacks that are both efficient and effective, or sparse, in the sense that it sensibly reduces the number of resources used. However, to the best of our knowledge, there is a lack of work that studies the impact of distance on these methods. In this paper, we set out to fix this.

3 Data

The data collected for the case study consists of a subset of the *Event Dataset* described by Albanese et al. in [4]. This dataset consists of a pool of 328.452 tweets in the English language from January 20th, 2021, retrieved through *Twitter Streaming API*. These tweets are specifically related to US President Biden's inauguration day. To gather the relevant tweets, several queries were used, including *Biden* (69.45%), *joebiden* (21.75%), *kamala-harris* (4.74%), and *inauguration2021* (4.04%).

To ensure the uniqueness of tweet labels, the authors removed from the pool any tweets that were retrieved by more than one query. Additionally, they pre-processed the remaining tweets by lower-casing all text and removing stop words.

In the case study, a directed graph $G = (V, E)$ was constructed, where each node v represents a Twitter user, while the edges represent retweet relationships between users [5]. Specifically, given two nodes in the graph $u, v \in V$, the directed edge (u, v) exists if u retweeted a tweet written by v. The entire network contains a total of 10.433 nodes and 29.626 edges.

The graph structure is represented through two matrices. The **adjacency matrix** is a $2 \times |E|$ matrix that defines the set of the edges in the graph. Each column contains a pair (*source, destination*) that describes a directed edge. On the other hand, given a set of node features F, the **feature matrix** is a matrix $|V| \times |F|$ matrix that represents node attributes and characteristics. In this case study, it has been generated through *laplacian positional encoding*. Each row is composed by 16 *Laplacian eigenvectors* [11], assigning positional information to every node.

4 Methods

4.1 Attack Setup

In order to perform the white-box attack, a GNN-based link prediction system was trained in a transductive setting [52], using 90% of existing edges as training set and the remaining 10% as test set, achieving an accuracy of 80%. Its model architecture consists of two stacked *convolutional layers* with respectively 128 and 64 hidden units, chained by a ReLU activation function. These convolutional layers are then followed by a two-layer MLP [43] with a ReLU activation function. Finally, the training process utilizes the Binary Cross-Entropy loss function and the Adam optimizer with a 0.01 learning rate. This is the model that will be later utilized for the link prediction task.

The experiment involves testing SAVAGE on the graph described in the previous section and employing the GNN link prediction model for the link prediction task. This task is performed on 6 different sets of pairs, each of them composed of 50 disconnected *source − target* node pairs, to discover if the directed links between them will be predicted.

The pairs in each set have the same distance length, and each set varies from the other for the pair distance in the range $[3, 8]$. This will show how the distance can affect the SAVAGE framework.

For this experiment, the maximum number of additional vicious nodes that can be introduced into the graph has been set to 50. The GNN model is iterated 150 times for each pair, adjusting the weights at every iteration. Similar to previous works [50], the constants β and γ that weight the loss function of the perturbed graph have been fixed to 0.1.

The link prediction model will predict the links on the original graph and on the perturbed one, showing the performance improvement.

4.2 Evaluation Metrics

In evaluating the performance of SAVAGE, several metrics are used and compared to the link prediction task on the unperturbed graph:

- **Kullback-Leibler Divergence**: It measures the difference between probability distributions [31]. In this context, it quantifies the node's degree distribution shift before and after the adversarial attack. This metric provides insights into how the generated perturbation affects the overall structure of the graph. Since this case study involves directed graphs, the KL divergence

is computed separately for both the node's indegree distribution (KI) and the outdegree distribution (KO). A higher value of the KL divergence corresponds to a stronger impact on the network topology, suggesting a more significant perturbation caused by the adversarial attack. Consequently, a lower KL divergence is desirable as it indicates less change in the graph's structure.
- **Mean Added Nodes** (AN): This metric evaluates the resource utilization and efficiency of the attack by measuring the number of nodes that are added to the perturbed graph. A lower value indicates better resource optimization and lower attack complexity, as it implies that fewer additional nodes were required to achieve the desired perturbation.
- **Initial and Final Prediction** (AP): These metrics measure the link prediction scores for a target link respectively before and after the attack. Each source-target pair is assigned a prediction score in the range $[0, 1]$, representing the likelihood of the link's existence. Given the link prediction model, the initial mean prediction indicates the actual prediction scores on the original graph, while the final mean prediction captures the values predicted on the perturbed graph. Comparing these metrics allows us to evaluate the effectiveness of the attack in an unbiased manner.
- **Attack Rate** (AR): Given the computed prediction score of a link, a threshold is used to determine whether the link is predicted to exist or not. If the predicted value is higher than the threshold, the attack is considered successful. The attack rate for a set of pairs is calculated as the proportion of successful attacks to the total number of attempted attacks:

$$AttackRate = \frac{\#successful\ attacks}{\#attacks}. \quad (1)$$

It is important to acknowledge that despite being strictly related to the final prediction metric, the attack rate can fluctuate depending on the choice for the threshold: a higher threshold leads to more accurate predictions, but it may result in a lower attack rate. In this case study, the threshold is set to 0.6, meaning that an attack is considered successful if the predicted probability of the target link is equal to or greater than 60%.

5 Results

5.1 SAVAGE White-Box Adversarial Attack

Firstly, a practical SAVAGE white-box adversarial attack was performed on the link prediction model over the Twitter user-retweet graph, and the distance-based sample sets mentioned earlier were tested.

Considering an aggregate result on all the 300 pairs at various distances in the range [3, 8], the SAVAGE attack achieved an attack rate of 61, 3%. The average final prediction was 0.60, compared to the 0.48 of the average initial prediction. That means a 25% of improvement at the cost of an average of only 6.77 vicious added nodes, which implies a 0.06% overhead in terms of nodes in the network.

The KL divergence was computed for both the indegree and outdegree distributions. The average KL divergence for the in-degree distribution (KL IN) was 1.1×10^{-21}, while for the outdegree distribution (KL OUT), it was 4.1×10^{-4}. Although the ideal value would be 0, these values are still very low, indicating a minimal impact on the network structure.

Upon closer and individual analysis of the results from the different distance-based sample sets, as depicted in Fig. 1b and 1a, it is observed that the attack exhibits better performance at shorter distances, even though the trend does not follow a linear pattern. Specifically, at a distance of 3, there is a peak final prediction score of 0.69, resulting in an attack rate of 70%. This represents an improvement in prediction accuracy of nearly 40% compared to the unperturbed graph, where the prediction score was 0.50.

On the other hand, the analysis of *KL divergence* and *mean added nodes* did not reveal any significant trend or correlation with the source-target distance. Thus, these metrics were considered negligible in this particular analysis.

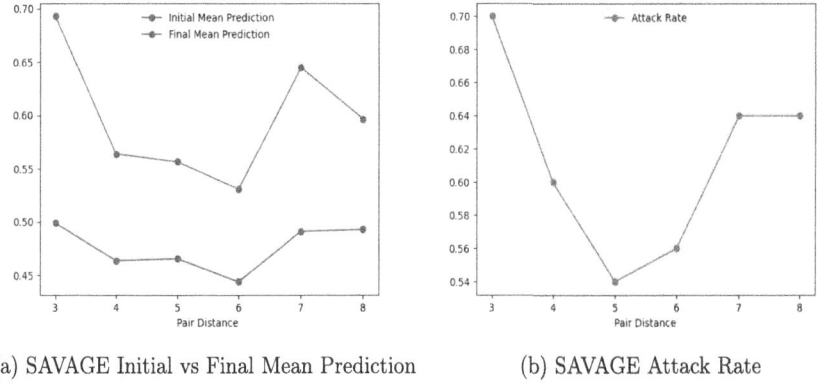

(a) SAVAGE Initial vs Final Mean Prediction   (b) SAVAGE Attack Rate

Fig. 1. SAVAGE Results on Distance-based Sample Sets

To provide robustness and reliability of results, confidence intervals were computed with a confidence level of 95%, as it is shown in Table 1. These intervals were computed using the bootstrap technique, which relies on the procedure of resampling. Starting from a data sample, a new sample of the same size is simulated considering every original value with replacement [44].

In Fig. 2, the bootstrap distributions of the final predictions are depicted. These distributions represent the ranges of possible final prediction values for each distance-based set that can be obtained through the bootstrap technique and provide a visual representation of the variability in the results.

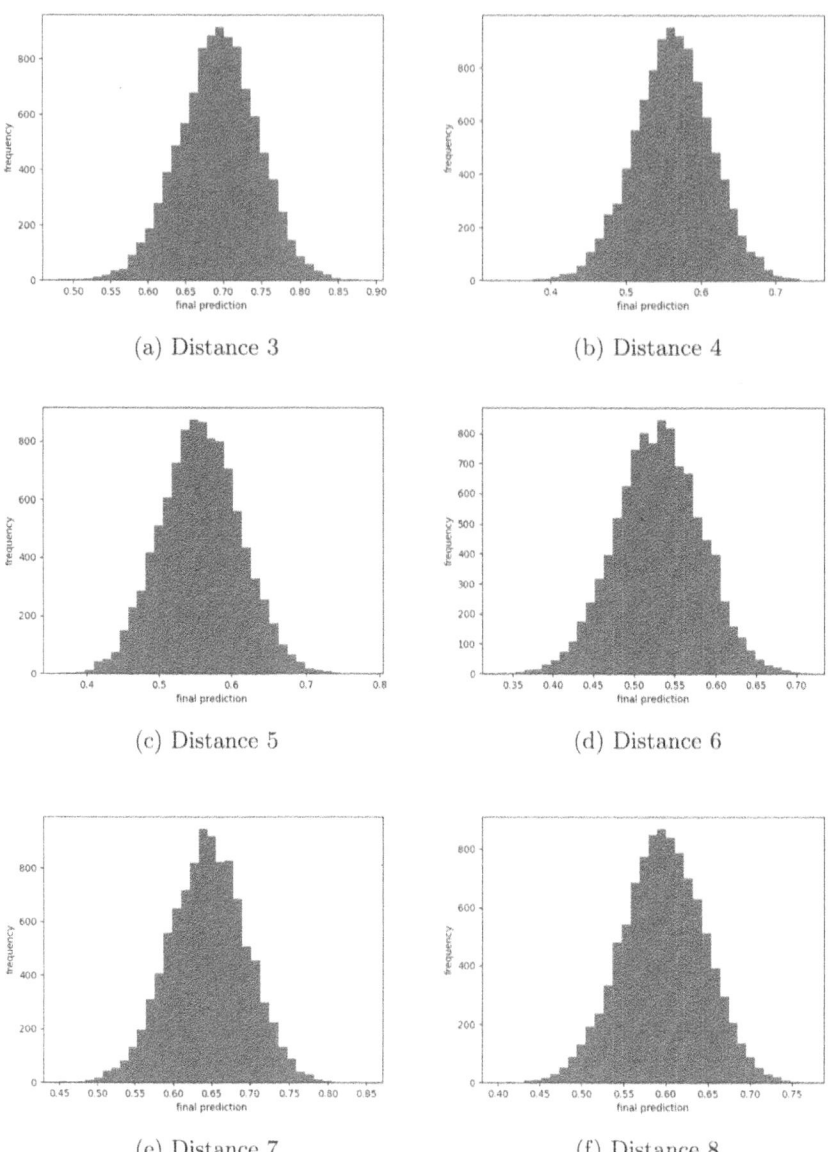

Fig. 2. Bootstrap distributions for final predictions

Table 1. SAVAGE 95% confidence intervals for final predictions

Distance	Predictions
3	0.684 ± 0.101
4	0.560 ± 0.101
5	0.554 ± 0.106
6	0.531 ± 0.104
7	0.640 ± 0.100
8	0.590 ± 0.096

5.2 Comparison with Other Adversarial Attacks

Different adversarial attacks were conducted on the same node pairs sample sets in order to compare SAVAGE performance with other attack frameworks. Initially, an adapted version of the Iterative Gradient Attack method was considered [14]. This adapted version, named **AIGA** (Adapted IGA), was specifically tailored to fit within the framework of this study. The IGA method has been recognized as one of the top-performing adversarial attacks on link prediction. Thus, by adapting it and testing it in this context, a more comprehensive understanding of the effectiveness of SAVAGE in relation to state-of-the-art attack methods is provided.

Moreover, a simple random baseline (RAND) was included in the evaluation. This approach allows for the addition/removal of connections in the graph and the activation/deactivation of vicious nodes with a probability p, which determines the strength of the attacker. In this case study, two random baselines were considered: **RAND-L**, with a low probability $p_L = 0.25$, and **RAND-H**, with a high probability $p_H = 0.75$.

Although random baselines may not achieve state-of-the-art performance due to their reliance on randomness, they provide an additional element of comparison in the evaluation process.

Figures at 3 provide a visual representation of the evaluation metrics for all the tested methods. The obtained results confirm the superiority of SAVAGE over the other attack frameworks in terms of final prediction (AP) and attack rate (AR) across all distances. It is worth noting that at a distance of 8, SAVAGE achieves a mean prediction score that is more than doubled compared to AIGA. This shows that SAVAGE not only performs better at shorter distances but also outclasses the state-of-the-art methods at greater distances. SAVAGE showcased its exceptional resource utilization as well. AIGA employed the maximum number of available additional nodes (AN) for every sample set, indicating low resource optimization. On the contrary, SAVAGE demonstrated a minimum usage of additional malicious resources, surpassing even RAND-L (which adds vicious nodes with a low probability).

Table 2. Results comparison between the state-of-the-art methods. Models were evaluated in terms of final prediction (AP), attack rate (AR), mean added nodes (AN), and KL-Divergence of the node's indegree distribution (KI) and outdegree distribution (KO).

Method	Distance 3					Distance 4				
	AP ↑	AR ↑	AN ↓	KI ↓	KO ↓	AP ↑	AR ↑	AN ↓	KI ↓	KO ↓
SAVAGE	**0.69**	**0.70**	8.62	$3 \cdot 10^{-21}$	$13 \cdot 10^{-4}$	**0.56**	**0.60**	3.98	0.00	0.00
AIGA	0.59	0.60	50	0.00	0.00	0.45	0.44	50	0.00	0.00
RAND-L	0.12	0.1	12.9	$2 \cdot 10^{17}$	$5 \cdot 10^{-3}$	0.15	0.06	12.9	$2 \cdot 10^{17}$	$5 \cdot 10^3$
RAND-H	0.10	0.08	37.54	10^{-17}	$5 \cdot 10^3$	0.12	0.06	37.54	10^{-17}	$5 \cdot 10^{-3}$

	Distance 5					Distance 6				
Method	AP ↑	AR ↑	AN ↓	KI ↓	KO ↓	AP ↑	AR ↑	AN ↓	KI ↓	KO ↓
SAVAGE	**0.56**	**0.54**	9.72	$3 \cdot 10^{-21}$	$12 \cdot 10^{-4}$	**0.53**	**0.56**	7.28	0.00	0.00
AIGA	0.53	0.52	50	0.00	0.00	0.47	0.48	50	0.00	0.00
RAND-L	0.07	0.02	12.9	$2 \cdot 10^{17}$	$5 \cdot 10^{-3}$	0.13	0.08	12.9	$2 \cdot 10^{17}$	$5 \cdot 10^{-3}$
RAND-H	0.06	0.02	37.54	10^{-17}	$5 \cdot 10^{-3}$	0.10	0.04	37.54	10^{-17}	$5 \cdot 10^{-3}$

	Distance 7					Distance 8				
Method	AP ↑	AR ↑	AN ↓	KI ↓	KO ↓	AP ↑	AR ↑	AN ↓	KI ↓	KO ↓
SAVAGE	**0.64**	**0.64**	7.38	0.00	0.00	**0.60**	**0.64**	3.68	0.00	0.00
AIGA	0.51	0.52	50	0.00	0.00	0.29	0.26	50	0.00	0.00
RAND-L	0.05	0.00	12.9	$2 \cdot 10^{17}$	$5 \cdot 10^{-3}$	0.16	0.16	12.9	$2 \cdot 10^{17}$	$5 \cdot 10^{-3}$
RAND-H	0.04	0.00	37.54	10^{-17}	$5 \cdot 10^{-3}$	0.13	0.12	37.54	10^{-17}	$5 \cdot 10^{-3}$

We have observed that SAVAGE consistently outperforms its competitors, particularly when the distance between nodes is very large. This performance advantage is primarily attributed to SAVAGE's ability to overcome proximity bias, recognizing that distant nodes are more important in certain scenarios. Proximity bias often leads other models to overly focus on nearby nodes, which can obscure critical connections involving distant nodes.

SAVAGE's sparse nature allows it to selectively ignore unnecessary nodes, thereby highlighting the most relevant paths and connections. This efficiency in node utilization means that SAVAGE can maintain a clear view of the network's structure, even as the distance between nodes increases. This characteristic is especially beneficial in applications such as social media friend/follower suggestions, where it is crucial to identify influential nodes and their connections despite their distance from the source node.

One minor weakness of SAVAGE in the evaluation is observed in relation to KL-Divergence (KI and KO): despite performing well, exhibiting small degree distribution shifts in the perturbed graph, AIGA achieved better performance at shorter distances.

For a more detailed overview of the results, Table 2 provides comprehensive information on all the evaluation metrics.

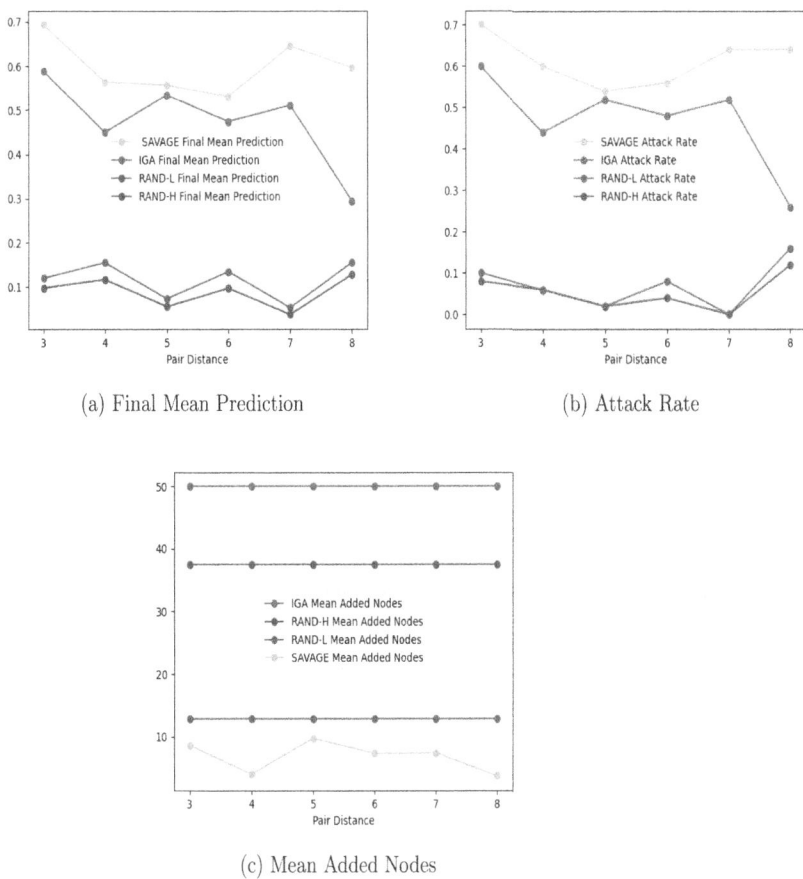

Fig. 3. SAVAGE vs IGA vs RAND-L vs RAND-H Comparison

6 Conclusions

The study analyzed the behavior of SAVAGE in relation to network topology and path length between source and target nodes. Results showcased the performance of the SAVAGE in comparison with the other methods. In particular, it confirms our initial hypothesis that SAVAGE is particularly good in large pair distances scenarios, which is particularly useful in real-world cases such as social media friend/follower suggestions where the source user (the attacker) is not close to the target user.

Future work could focus on SAVAGE transferability and analyze the effectiveness of transferring adversarial perturbations generated by SAVAGE to deceive other link prediction systems operating under a black-box setting. This could lead to the design of link prediction models that can effectively defend against such attacks. Future work would also validate the results in more diverse scenarios with other datasets.

Acknowledgements. This work was supported by PNRR MUR projects PE0000013-FAIR, SERICS (PE00000014), IR0000013-SoBigData.it.

References

1. Adamic, L.A., Adar, E.: Friends and neighbors on the web. Soc. Netw. **25**(3), 211–230 (2003). https://doi.org/10.1016/S0378-8733(03)00009-1, https://www.sciencedirect.com/science/article/pii/S0378873303000091
2. Agostinelli, A., et al.: Musiclm: generating music from text (2023)
3. Al Hasan, M., Chaoji, V., Salem, S., Zaki, M.: Link prediction using supervised learning. In: Proceedings of of SDM 2006: Workshop on Link Analysis, Counter-Terrorism and Security, vol. 30, pp. 798–805 (2006)
4. Albanese, F., Feuerstein, E.: Improved topic modeling in twitter through community pooling. In: Lecroq, T., Touzet, H. (eds.) String Processing and Information Retrieval, SPIRE 2021, LNCS, vol .12944, pp. 209–216. Springer, Cham (2021). https://doi.org/10.1007/978-3-030-86692-1_17
5. Albanese, F., Feuerstein, E., Lombardi, L., Balenzuela, P.: Characterizing community-changing users using text mining and graph machine learning on twitter (2021)
6. Bacciu, A., Cuconasu, F., Siciliano, F., Silvestri, F., Tonellotto, N., Trappolini, G.: RRAML: reinforced retrieval augmented machine learning. In: Basili, R., Lembo, D., Limongelli, C., Orlandini, A. (eds.) Proceedings of the Discussion Papers - 22nd International Conference of the Italian Association for Artificial Intelligence (AIxIA 2023 DP) co-located with 22nd International Conference of the Italian Association for Artificial Intelligence (AIxIA 2023), Rome, Italy, November 6-9, 2023. CEUR Workshop Proceedings, vol. 3537, pp. 29–37. CEUR-WS.org (2023). https://ceur-ws.org/Vol-3537/paper4.pdf
7. Bacciu, A., Trappolini, G., Santilli, A., Rodolà, E., Silvestri, F.: Fauno: the Italian large language model that will leave you senza parole! In: Nardini, F.M., Tonellotto, N., Faggioli, G., Ferrara, A. (eds.) Proceedings of the 13th Italian Information Retrieval Workshop (IIR 2023), Pisa, Italy, 8–9 June 2023. CEUR Workshop Proceedings, vol. 3448, pp. 9–17. CEUR-WS.org (2023). https://ceur-ws.org/Vol-3448/paper-24.pdf
8. Barabási, A.L., Albert, R.: Emergence of scaling in random networks. ArXiv preprint **abs/10.1126** (2010). https://arxiv.org/abs/10.1126
9. Barnabò, G., et al.: CycleDRUMS: automatic drum arrangement for bass lines using CycleGAN. Disc. Artif. Intell. **3**(1), 4 (2023)
10. Basile, P., Musacchio, E., Polignano, M., Siciliani, L., Fiameni, G., Semeraro, G.: LLaMAntino: LLaMA 2 models for effective text generation in Italian language (2023)
11. Belkin, M., Niyogi, P.: Laplacian eigenmaps for dimensionality reduction and data representation. Neural Comput. **15**, 1373–1396 (2003)
12. Brin, S., Page, L.: The anatomy of a large-scale hypertextual web search engine. Comput. Netw. ISDN Syst. **30**(1), 107–117 (1998). https://doi.org/10.1016/S0169-7552(98)00110-X, https://www.sciencedirect.com/science/article/pii/S016975529800110X

13. Brown, T., et al.: Language models are few-shot learners. Adv. Neural. Inf. Process. Syst. **33**, 1877–1901 (2020)
14. Chen, J., Lin, X., Shi, Z., Liu, Y.: Link prediction adversarial attack via iterative gradient attack. IEEE Trans. Comput. Soc. Syst. **7**(4), 1081–1094 (2020)
15. Chen, J., Wu, Y., Xu, X., Chen, Y., Zheng, H., Xuan, Q.: Fast gradient attack on network embedding. arXiv preprint arXiv:1809.02797 (2018)
16. Chen, Z., Silvestri, F., Tolomei, G., Wang, J., Zhu, H., Ahn, H.: Explain the explainer: interpreting model-agnostic counterfactual explanations of a deep reinforcement learning agent. IEEE Trans. Artif. Intell. 1–15 (2022). https://doi.org/10.1109/TAI.2022.3223892
17. Chen, Z., et al.: GREASE: generate factual and counterfactual explanations for GNN-based recommendations. ArXiv preprint **abs/2208.04222** (2022). https://arxiv.org/abs/2208.04222
18. Chen, Z., Silvestri, F., Wang, J., Zhang, Y., Tolomei, G.: The dark side of explanations: poisoning recommender systems with counterfactual examples. In: Proceedings of SIGIR 2023, pp. 2426–2430. ACM (2023)
19. Chen, Z., Silvestri, F., Wang, J., Zhu, H., Ahn, H., Tolomei, G.: ReLAX: reinforcement learning agent explainer for arbitrary predictive models. In: Proceedings of CIKM 2022, pp. 252–261. ACM (2022)
20. Copet, J., Kreuk, F., Gat, I., Remez, T., Kant, D., Synnaeve, G., Adi, Y., Défossez, A.: Simple and controllable music generation (2024)
21. Cui, Y., Yang, Z., Yao, X.: Efficient and effective text encoding for Chinese llama and alpaca. arXiv preprint arXiv:2304.08177 (2023). https://arxiv.org/abs/2304.08177
22. Dai, E., et al.: A Comprehensive survey on trustworthy graph neural networks: privacy, robustness, fairness, and explainability. ArXiv preprint **abs/2204.08570** (2022), https://arxiv.org/abs/2204.08570
23. Dai, J., Zhu, W., Luo, X.: A targeted universal attack on graph convolutional network by using fake nodes. Neural Process. Lett. **54**, 1–17 (2022). https://doi.org/10.1007/s11063-022-10764-2
24. Garrachonr: Llamados. https://github.com/Garrachonr/LlamaDos (2023)
25. Goodfellow, I.J., Shlens, J., Szegedy, C.: Explaining and harnessing adversarial examples. In: Proceedings of ICLR 2015 (2015). http://arxiv.org/abs/1412.6572
26. Grosse, K., Papernot, N., Manoharan, P., Backes, M., McDaniel, P.: Adversarial examples for malware detection. In: Foley, S.N., Gollmann, D., Snekkenes, E. (eds.) ESORICS 2017. LNCS, vol. 10493, pp. 62–79. Springer, Cham (2017). https://doi.org/10.1007/978-3-319-66399-9_4
27. Ho, J., Jain, A., Abbeel, P.: Denoising diffusion probabilistic models (2020)
28. Ho, J., Salimans, T., Gritsenko, A., Chan, W., Norouzi, M., Fleet, D.J.: Video diffusion models (2022)
29. Jeh, G., Widom, J.: SimRank: a measure of structural-context similarity. In: Proceedings of KDD 2002, pp. 538–543. ACM (2002). https://doi.org/10.1145/775047.775126
30. Jiang, A.Q., et al.: Mistral 7b (2023)
31. Joyce, J.: Kullback-leibler divergence, pp. 720–722 (2011)
32. jphme: Llama-2-13b-chat-German. https://huggingface.co/jphme/Llama-2-13b-chat-german (2023)
33. Katz, L.: A new status index derived from sociometric analysis. Psychometrika **18**(1), 39–43 (1953)
34. Ke, Z., Kong, W., Li, C., Zhang, M., Mei, Q., Bendersky, M.: Bridging the preference gap between retrievers and llms. arXiv preprint arXiv:2401.06954 (2024)

35. Kumar, A., Singh, S.S., Singh, K., Biswas, B.: Link prediction techniques, applications, and performance: a survey. Phys. A **553**, 124289 (2020)
36. Li, B., Xia, Y., Xie, S., Wu, L., Qin, T.: Distance-enhanced graph neural network for link prediction. In: Proceedings of ICML 2021: Workshop on Computational Biology (2021)
37. Liben-Nowell, D., Kleinberg, J.: The link prediction problem for social networks. In: Proceedings of CIKM 2003, pp. 556–559. ACM (2003). https://doi.org/10.1145/956863.956972
38. Lin, W., Ji, S., Li, B.: Adversarial attacks on link prediction algorithms based on graph neural networks. In: Proceedings Of The 15th ACM Asia Conference On Computer And Communications Security, pp. 370–380 (2020)
39. Lü, L., Zhou, T.: Link prediction in Complex networks: a survey. Phys. A **390**(6), 1150–1170 (2011)
40. Lucic, A., Ter Hoeve, M.A., Tolomei, G., De Rijke, M., Silvestri, F.: CF-GNNExplainer: counterfactual explanations for graph neural networks. In: Proceedings of AISTATS 2022, pp. 4499–4511. PMLR (2022)
41. Martínez, V., Berzal, F., Cubero, J.C.: A survey of link prediction in complex networks. ACM Comput. Surv. **49**(4), 1–33 (2016) https://doi.org/10.1145/3012704
42. Newman, M.E.: Clustering and preferential attachment in growing networks. Phys. Rev. E **64**(2), 025102 (2001)
43. Raj, P., David, P.: The Digital Twin Paradigm for Smarter Systems and Environments: the Industry use Cases. Academic Press, Cambridge (2020)
44. Ramachandran, K., Tsokos, C.: Mathematical Statistics with Applications in R. Academic Press, Cambridge (2020)
45. Sun, Y., Wang, S., Tang, X., Hsieh, T.Y., Honavar, V.: Adversarial attacks on graph neural networks via node injections: a hierarchical reinforcement learning approach. In: Proceedings of TheWebConf 2020, pp. 673–683 (2020)
46. Tolomei, G., Campagnano, C., Silvestri, F., Trappolini, G.: Prompt-to-os (P2OS): revolutionizing operating systems and human-computer interaction with integrated AI generative models. In: 5th IEEE International Conference on Cognitive Machine Intelligence, CogMI 2023, Atlanta, GA, USA, 1–4 November 2023, pp. 128–134. IEEE (2023). https://doi.org/10.1109/COGMI58952.2023.00027
47. Tolomei, G., Silvestri, F.: Generating actionable interpretations from ensembles of decision trees. IEEE Trans. Knowl. Data Eng. **33**(4), 1540–1553 (2021)
48. Tolomei, G., Silvestri, F., Haines, A., Lalmas, M.: Interpretable predictions of tree-based ensembles via actionable feature tweaking. In: Proceedings of KDD 2017, pp. 465–474. ACM (2017)
49. Touvron, H., et al.: Llama: open and efficient foundation language models (2023)
50. Trappolini, G., Maiorca, V., Severino, S., Rodola, E., Silvestri, F., Tolomei, G.: Sparse vicious attacks on graph neural networks. IEEE Trans. Artif. Intell. **5**, 2293–2303 (2023)
51. Trappolini, G., Santilli, A., Rodolà, E., Halevy, A., Silvestri, F.: Multimodal neural databases. In: Proceedings of the 46th International ACM SIGIR Conference on Research and Development in Information Retrieval, pp. 2619–2628 (2023)
52. Tripodi, R., Pelillo, M.: Transductive Learning Games for Word Sense Disambiguation, pp. 109–128 (2017)
53. Wang, J., Luo, M., Suya, F., Li, J., Yang, Z., Zheng, Q.: Scalable Attack on Graph Data by Injecting Vicious Nodes. Data Min. Knowl. Disc. **34**(5), 1363–1389 (2020)
54. Wang, X., Cheng, M., Eaton, J., Hsieh, C.J., Wu, F.: Attack graph convolutional networks by adding fake nodes. ArXiv preprint **abs/1810.10751** (2018), https://arxiv.org/abs/1810.10751

55. Wu, Z., Pan, S., Chen, F., Long, G., Zhang, C., Philip, S.Y.: A comprehensive survey on graph neural networks. IEEE Trans. Neural Netw. Learn. Syst. **32**(1), 4–24 (2020)
56. Xie, T., et al.: Osworld: benchmarking multimodal agents for open-ended tasks in real computer environments (2024)
57. Zhang, M., Chen, Y.: Link prediction based on graph neural networks. In: Proceedings of NeurIPS 2018, pp. 5171–5181 (2018), https://proceedings.neurips.cc/paper/2018/hash/53f0d7c537d99b3824f0f99d62ea2428-Abstract.html
58. Zhou, T., Lü, L., Zhang, Y.C.: Predicting missing links via local information. Eur. Phys. J. B **71**(4), 623–630 (2009)
59. Zügner, D., Akbarnejad, A., Günnemann, S.: adversarial attacks on neural networks for graph data. In: Proceedings of of KDD 2018, pp. 2847–2856. ACM (2018). https://doi.org/10.1145/3219819.3220078

Author Index

A
Albanese, Federico 73

B
Bufi, Salvatore 41

C
Chen, Hung-Hsuan 60
Cohen, Nachshon 25

D
De Filippis, Giovanni M. 11
Di Noia, Tommaso 41
Di Sciascio, Eugenio 41

F
Ferrara, Antonio 41
Fetahu, Besnik 25

G
Gautam, Sushant 1

H
Halvorsen, Pål 1
Haramaty, Elad 25

L
Lewin-Eytan, Liane 25

M
Malitesta, Daniele 41
Malmasi, Shervin 25
Mancino, Alberto Carlo Maria 41
Midoglu, Cise 1

R
Riegler, Michael A. 1
Rinaldi, Antonio M. 11
Rokhlenko, Oleg 25
Russo, Cristiano 11

S
Scarlino, Lorenzo 73
Sepasdar, Zahra 1
Silvestri, Fabrizio 73

T
Tommasino, Cristian 11
Trappolini, Giovanni 73

W
Wang, Pei-Syuan 60

SPRINGER NATURE

GPSR Compliance

The European Union's (EU) General Product Safety Regulation (GPSR) is a set of rules that requires consumer products to be safe and our obligations to ensure this.

If you have any concerns about our products, you can contact us on ProductSafety@springernature.com

In case Publisher is established outside the EU, the EU authorized representative is:

Springer Nature Customer Service Center GmbH
Europaplatz 3
69115 Heidelberg, Germany

The manufacturer's authorised representative in the EU is Springer Nature Customer Service Centre GmbH, Europaplatz 3, 69115 Heidelberg, Germany. If you have any concerns regarding our products, please contact ProductSafety@springernature.com

Printed and bound by CPI Group (UK) Ltd, Croydon, CR0 4YY

25/03/2026

02078190-0011